Rapid Prototyping Casebook

Rapid Prototyping Casebook

Edited by

J A McDonald, C J Ryall, and D I Wimpenny

**Professional
Engineering
Publishing**

Professional Engineering Publishing Limited
London and Bury St Edmunds, UK

First Published 2001

ISBN 1 86058 076 9

The terms 'ThermoJet' and 'SLA' (stereolithography), used in this book are registered trademarks of the company 3D Systems, Valencia, California, www.3dsystems.com

A CIP catalogue record for this book is available from the British Library.

Printed and bound in Great Britain by The Cromwell Press, Trowbridge, Wiltshire, UK

Further Titles of Interest

Rapid Prototyping and Tooling Research	Conference publication G Bennett (Ed)	ISBN 0 85298 982 2
Developments in Rapid Prototyping and Tooling	Conference publication G Bennett (Ed)	ISBN 1 86058 048 3
IMechE's Engineer's Data Book (Second Edition)	C Matthews	ISBN 1 86058 248 6
A Guide to Presenting Technical Information – Effective Graphic Communication	C Matthews	ISBN 1 86058 249 4
Advanced Manufacturing Processes, Systems, and Technologies (AMPST '99)	Conference publication M K Khan, Y T Abdul-Hamid, C S Wright, and R Whalley (Eds)	ISBN 1 86058 230 3
Advances in Manufacturing Technology XII	Conference publication R W Baines, A Taleb-Bendiab, and Z Zhao (Eds)	ISBN 1 86058 172 2
Advances in Manufacturing Technology XIII	Conference publication A N Bramley, A R Mileham, L B Newnes, and G W Owen (Eds)	ISBN 1 86058 227 3
16th International Conference on Computer Aided Production Engineering – CAPE 2000	IMechE Conference publication J A McGeough (Ed)	ISBN 1 86058 263 X

For the full range of titles published by Professional Engineering Publishing contact:

Sales Department
Professional Engineering Publishing Limited
Northgate Avenue
Bury St Edmunds
Suffolk
IP32 6BW
UK

Tel: +44 (0)1284 724384
Fax: +44 (0)1284 718692
Website: www.pepublishing.com

Contents

Section 1

Product Design and Development

Overview of product design and development

Julia McDonald

Industry and consumers have, over the years, come to expect higher levels of product quality and reliability. To be globally competitive, companies are under increasing pressure to introduce new products and processes to the marketplace within shorter lead times and at the same time offer increased quality and performance.

It is vital to ensure that components fit together to form a product that is suitable for the working environment at an early stage, to avoid costly and time-consuming remedial work. If manufacturing can also be addressed early in the design process, production and quality control problems can be avoided. Savings of as little as 1–2 per cent within the overall process (manufacturing, assembly, packaging, distribution, etc.) can be the competitive advantage that differentiates success from failure. Moreover, the implications of bringing a product to market late should not be underestimated, particularly in the rapidly evolving sectors, such as telecommunications, for example. For a typical consumer product, management consultants McKinsey & Co., report that a 50 per cent increase in the budget for developing the product has less than a tenth of the financial implications of being six months late introducing it to the market (1). Shortcomings in the performance or reliability of products can also be disastrous for a company. Expenditure resulting from actual or perceived product failures from the time of distribution over the life of the product can become a major financial burden for the manufacturer. In addition, manufacturing organizations may find it impossible to prevent bad publicity, relating to the failure of one product, from tarnishing the business as a whole, perhaps undermining years of hard work.

If time, effort, and resources are to be used productively and rapid prototyping is to be utilized, then usable CAD data is a necessity. Peter Dicken discusses some of the important features with surface and solid modelling CAD systems, their relative merits and issues relating to the generation of a good STL file. TNO Institute of Technology, together with partners across Europe, has been investigating the application of CAD/CAM and rapid prototyping technology within the ceramics industry. In this paper the changes which traditional industries must make, in order to take full advantage of this new approach, are illustrated using a series of case studies. The next paper is also centered on the ceramics industry and describes work carried out within a UK funded development programme. Vinesh Raja describes the application of reverse engineering for the development of a ceramic floor tile by capturing and manipulating geometric data from nature. Using this approach, the design process was reduced from 49 days to just five days. Steve May-Russell and Martin Smith, Smallfry Industrial Design, show how more progressive companies can rethink the whole of the design process to incorporate a whole range of new methods, including rapid prototyping. This new approach is illustrated using case studies based on the design of fashion based consumer products.

Rapid prototyping in not just restricted to aesthetic and ergonomic design issues. The final two papers of this chapter explore the use of rapid prototyping techniques to validate and optimize the performance of products in respect of both stress and flow characteristics. In the first of these two papers, Geoff Calvert describes the application of empirical stress analysis methods with RP models for the development of an automotive gearbox and shows that this approach can complement computational methods, such as finite element analysis. In the last paper of this Section, Chris Driver describes the use of flow analysis, with stereolithography models, to optimize the performance of a new engine design. Although both stress and flow analysis methods have been pioneered in the automotive industry they can provide an extremely useful design tool for a wide range of products.

In conclusion, the product design and development process has many disciplines and stages to enable a working solution to be finalized. The techniques and information covered in this Section relate to issues from CAD implementation, to physical testing, and illustrate methods to assist in the product design process. These should be seen as additional tools available to designers and engineers, and not as the answer to every solution. Rapid prototyping technologies have allowed greater levels of product validation to be achieved in shorter time scales, enabling products to meet customer expectations. However, any technique needs to be applied appropriately and in context with traditional and other advanced technologies.

REFERENCE

1   **Reeve, R.** (1992) Profiting from teamwork. *Manufacturing Breakthrough*, Jan/Feb issue.

Computer-aided design and rapid prototyping

Peter Dickin
Marketing Manager, Delcam UK, Small Heath, Birmingham, UK

All rapid prototyping (RP) projects require, as the first stage, the creation of a three-dimensional (3D) computer model with a computer-aided design (CAD) system. The CAD software provides the medium for defining geometry and verifying aesthetic and dimensional characteristics at the modelling stage. The completed CAD model is the central definition of component geometry needed for physical prototype manufacture. Like most other processes involving computers, rapid prototyping equipment follows the 'garbage in – garbage out' rule faithfully. An important part of any rapid prototyping operation is, therefore, to ensure that the CAD model is of high quality.

Quality can be divided into two aspects. Firstly, there is a need to ensure that the CAD model accurately reflects the component or product that will finally be made. In the early days of rapid prototyping, this aspect was related solely to the ability of the model to capture the design intent accurately. More recently, other factors have been included within this definition, in particular related to the manufacturability of the design. After all, original and imaginative designs are of little practical use if there is no process capable of mass producing them.

The second aspect of quality is the mathematical accuracy and completeness of the model. Rapid prototyping equipment is much less forgiving than, for example, milling machines, so a more mathematically complete model is required to produce samples successfully.

1   SURFACE MODELLING AND SOLID MODELLING

Computer-aided design systems used to be divided into two principal types – those based on solid modelling and those focusing on surface modelling. More recently, systems combining both of these approaches have been introduced – the hybrid modellers, such as Delcam's PowerSHAPE CAD system. Before comparing the two modelling techniques, two definitions are required:

- *complicated parts* are defined as those built up from a large number of basic geometric shapes, that is planes, cylinders, cones, etc.;
- *complex parts* are defined as those containing complex shapes, such as surfaces with multiple curves.

An example of a complicated part would, with these definitions, be a knife-block, since its design is made up of a large number of standard geometric shapes. In contrast, a spoon would be considered a complex model, even though it contains fewer surfaces, since many of those surfaces are doubly curved. It must be realized that it is a more difficult mathematical challenge to model a complex shape, even if it has only a few surfaces, than a complicated shape with a large number of surfaces.

1.1   Surface modelling

Surface modellers usually operate through a combination of two methods. The initial surfaces may be defined over a network of curves or by creating a number of cross-sections along the length of the object and then producing a smooth surface along the outside of the cross-sections. Alternatively, wire-frame modelling may be used first to define the basic geometry. In simple terms, wire-frame modelling is an extension of two-dimensional drafting to three dimensions. It allows the user to build up a wire-frame representation of the object by joining points in space with lines, arcs, or curves. Surfaces may then be placed over the wire-frame geometry to create the surface model.

Once the surfaces are created, the user may select and move individual points, or groups of points, within them. The CAD program will then reshape the surface to incorporate the new position of the point or points. The effects of the change may be limited to a single surface or to a group of adjoining surfaces. This dynamic styling capability gives the ability to create any shape that the designer can imagine.

Surface modelling was used as the basis for most early 3D CAD systems. These early systems were much more complex than current CAD software. They involved a hierarchy of geometrical objects that the user had to know about – surfaces with internal control curves, each of which had control points plus tangent vectors or knots and weights; plus trim curves, each of which had its own points, etc. Each of these objects had its own set of edit commands that a user had to learn. Furthermore, there were extra commands involved in linking or stitching individual surfaces together in a smooth way to form the complete, continuous 'skin' over the model. Most of the early problems in using surface modellers with RP equipment came from difficulties with these stitching operations.

1.2   Solid modelling

Solid modellers also operate in one of two general ways. The model may be created by drawing a cross-section in two dimensions, which is either extruded or rotated to produce a solid. Alternatively, primitive geometric shapes can be produced (spheres, cylinders, cubes, etc.), which can be combined to produce models of parts.

When they were introduced, solid modellers immediately had the advantage of being easier to understand since they often only involve working with a single class of object – the solid. The solid, in its turn, can be manipulated by only three basic editing commands (at least conceptually) – the Boolean operations of subtraction, addition, and intersection. In addition, solid modellers, in most cases, don't allow individual surfaces to break apart, so commands to handle the linking of the surfaces of the skin are not necessary.

Modern solid modellers provide a longer list of edit operations than just the three Booleans, by combining them in various ways as 'features'. Even so, they are generally much easier to learn to use than surface modellers because they are dealing with many fewer types of object, using many fewer editing functions.

Also, the fact that, at the low, internal level, there are really only three basic edit operations means that it is possible to build associativity and parameterization into the modelling process. Thus, a set of lines can be defined always to have the same length or a set of arcs can be given the same radius. Any subsequent change to one of the lines or arcs would then automatically produce the same change in the other linked items. Detailed mathematical formulae can be built up to define the relationships between the various elements of the design so that altering one key parameter can alter the whole design. This ability to change a parameter and have the rest of the model update automatically makes solid modellers ideal for conceptual design where designers must try out various 'what-if?' scenarios.

The relative simplicity of solid modelling also makes it feasible to maintain a history of operations performed. Any changes to any of the historical operations can then be carried through the design process automatically. The parametric, history-based approach to modelling is of great benefit when designing large assemblies, since changes to one component can be reflected in any associated parts. For example, a change to the piston length could automatically modify the entire design of an engine. Clearly this capability can significantly reduce the modelling time, especially when a number of variants are required to the same basic assembly, for example, a series of engines with different capacities.

2   CAD SYSTEMS FOR RAPID PROTOTYPING

Rapid prototyping has traditionally been associated with solid modelling CAD systems rather than surface modelling programs. This is principally because the solid models produced are expected to have complete integrity, with no gaps that could cause problems for the rapid prototyping equipment. As explained above, surface modellers require the user to apply extra operations to ensure that the surfaces making up the model are stitched together correctly.

Also, surface modelling systems were generally developed with machining in mind. Gaps between surfaces were not critical provided that the size of the gaps was less than the diameter of the smallest cutter to be used in machining the part. Thus, unless the tolerances within the CAD system were modified, the resulting models could easily have gaps that would cripple rapid prototyping equipment.

Even so, despite their popularity for the development of RP models, it is becoming recognized that solid modellers are by no means perfect for every application. They can model the bulk of engineering components that merely need to perform a function. In such cases, the shapes are usually fairly simple and well within the capabilities of solid modelling. Also, the exact shape is not a key consideration for the overall design, so restrictions imposed by the software are not critical.

In contrast, solid modellers are not very satisfactory at handling complex geometries with free-form, sculptured surfaces, since they are based on prismatic shapes like cubes, spheres, and cones. Many real products cannot be modelled by these primitives alone. Curves used by surface modellers are more flexible and so more capable of modelling complex shapes. Classic examples include automotive styling, turbine blades, and airframe design. Such shapes are also essential for any product where aesthetics or ergonomics are an important consideration.

The second potential problem is that parametric modelling can be too formal. The use of history-based parametrics to generate a series of designs requires a formal approach to modelling. Anyone wishing to modify the model needs to understand how it was first constructed, while unplanned changes can be difficult if not impossible. This formality can be especially restrictive in the early, conceptual stages of the design processes. In contrast, the interactive modelling and styling possible with surface modellers makes design modification much more flexible. This flexibility can be a key factor where the appearance of the product is of greater importance than functional requirements.

A number of current design trends are also increasing the need for surface modelling. The first of these trends is the growing desire for 'organic' shapes, which is spreading from luxury goods to all areas of mechanical design. Product designers everywhere are trying to make their products more aesthetically pleasing. A typical example is home electronics, where shapes are moving from rectangular boxes to enclosures with complex, rounded surfaces. Similarly, complex packaging designs are spreading from luxury items, like perfumes and cosmetics, into many everyday goods.

The second trend is the recognition of the importance of ergonomics in product design. Human geometry is extremely complex with no simple shapes. Any product that touches the human body is ripe for surface modelling technology. Even when it is not essential for comfort, ergonomic design is a growing differentiator with many products.

Two manufacturing trends are also leading to more complex parts. Firstly, more powerful processing equipment is enabling increased use of component integration to replace a number of small parts with one larger component. This can have a significant effect on profitability since it allows assembly costs to be reduced. However, the shape of this single component will inevitably be more complex and, therefore, more likely to require surface modelling.

At the other end of the scale, miniaturization is also increasing complexity. In electrical appliances in particular, but also in other industries, products are being made smaller and smaller. Compare the size of a current mobile phone with one from only two years ago. The drive for miniaturization makes exact definition of shape more important. Restrictions imposed by inaccurate prototypes are no longer acceptable when working with assemblies that must be combined with tighter tolerances in smaller volumes.

Surface modellers are usually more flexible when it comes to creating a model with data from a number of different sources, such as existing CAD data or digitized data collected from earlier designs. The latter capability is becoming more important as companies realize that many 'new' parts are, in fact, variations on existing components. It is often quicker to digitize the existing part and limit the CAD work to the desired modifications, instead of completely recreating the part with CAD.

Finally, the increased experience with rapid prototyping has emphasized the need for accurate geometry to make the results meaningful. Companies are realizing the problems that result when models are just approximations with some details left out. By allowing the designer to generate complete, accurate CAD models, surface modellers enable RP models to be made that genuinely represent the part that will finally be manufactured. In contrast, systems without this level of design flexibility often produce models that need to have complex details, such as fillets, added by hand. This jeopardizes the value of any tests done on the models as well as losing the time advantages offered by rapid prototyping. In addition, the repeatability of these hand additions cannot be guaranteed on a series of models.

3   DESIGN FOR MANUFACTURE

The need to speed product development by breaking down the barriers between departments has been promoted for at least 30 years. However, the improved communications made possible by the Internet have made this philosophy much more achievable. With current technology, no individual or department is ever more than an e-mail away. It is, therefore, much easier for the designer to consult with his colleagues and ensure, for example, that his model does not include fillets that are too small to allow the material to flow successfully or that his new packaging design will fit on the company's existing filling lines.

Rapid prototyping can make a major contribution in this area, even though it will emphasize the need for the prototypes to represent the final component exactly. For example, prototypes can be produced for testing assembly lines and for training new operators at the same as the tooling is being manufactured for the 'real' components. Similarly, prototype tooling can be manufactured, either directly in the RP equipment or by casting from a prototype model. This tooling can then be used both to produce short runs of trial components and to test the flow of material in the tool and the ease with which the part can be removed. With this type of testing, designs should no longer be presented to the toolmaker or other manufacturing departments that are impossible to manufacture.

For this scenario to be possible, the CAD system must first be capable of producing a design as it will need to be made, with all draft surfaces and fillets included. It must also be flexible enough to be able to make any modifications that are required as a result of the tests.

4   REPAIR OF CAD MODELS AND STL FILES

Until now, this article has focused on the role of the CAD system creating an accurate, fully detailed model. The second stage is ensuring that this model can be converted into an STL file of the required quality to meet the demands of the RP equipment. As mentioned above, most solid modellers will produce a suitable model, while surface modellers can leave small, but critical gaps, between the component surfaces. However, even with solid modellers, there can still be problems with data that have been translated between alternative CAD systems, for example when the product designer and the rapid prototyping bureau use different software.

To overcome this problem, Delcam has added a 'make watertight' command within its PowerSHAPE software. This works by identifying surfaces where the trim boundaries are above the required tolerance and then re-intersecting the edges to give an exact match. Similar 'healing' operations are available within CAD software based on the ACIS and Parasolid modelling kernels.

A further quality process is required after the STL file has been created. To ensure a satisfactory prototype, the triangulated model in the files should obey certain rules; in particular the model should be closed with the vertex of each triangle common to two other triangles. These rules ensure that the triangles properly enclose the solid material in the part, so that slicing the STL model to produce the geometry of a closed laminar for the RP machine will always work. Delcam has developed an STL model verification and editing module called TriFIX, which ensures that models meeting these rules can be generated by repairing gaps in the model and correcting overlapping surfaces. TriFIX also allows a range of other common problems with STL files to be rectified. Duplicate triangles and nodes can be removed, so can triangles with less than a given height, intersecting triangles, and triangles that have more than one other triangle sharing an edge. Using this software can prevent unnecessary time and expense caused by trying to produce an RP model from a damaged or corrupt STL file.

4.1   Reverse engineering from models or prototypes

Almost as soon as any models or prototypes are created, there will be a need to modify them. These changes may be required for improved performance, for example to improve the gas flow in an automotive engine component. Equally they may be for aesthetic reasons, such as a styling modification to a design for packaging or a consumer product. These changes will frequently be very small, particularly towards the end of the cycle of prototyping and testing. It is often difficult and time consuming to reproduce such small modifications on the CAD model within the computer. An alternative approach is to digitize the region where the change has been made, and use the digitized data to update CAD models.

To speed this process, Delcam has developed a reverse engineering program, called CopyCAD, to allow quick and easy creation of CAD surfaces from digitized data. This accepts data from co-ordinate measuring machines, tracing machines, or laser scanners, and outputs surfaces that can be transferred into all major CAD systems.

Once the data have been loaded, a range of editing tools is available to remove excess data points and to create a more accurate model by offsetting for wall thickness, shrinkage, or measuring probe allowance. The edited data are then converted into a triangulated model. All of these initial stages can be carried out using a reverse engineering 'wizard', which automates the process and so makes the software very easy to use.

When creating surfaces, the user is given complete control over the selection of boundaries. This means that surfaces can be optimized for subsequent modelling and machining operations. After each set of boundary points has been chosen, the software automatically generates a smooth, multi-patch surface to the specified tolerance. Most importantly for those involved in RP, the software ensures that no gaps appear between the boundaries. Once the modified surfaces from the prototype have been created they can be combined with the main CAD model in the computer. The new model can then be used for another round of prototype manufacture and modification if required.

5   CONCLUSIONS

The choice of the most suitable CAD software is an important part of any RP operation. CAD developers like Delcam have responded to the demands of the rapid prototyping community by developing specific tools for its needs, alongside more general improvements in power, flexibility, and ease of use. A key advance has been the development of hybrid modelling software, combining the benefits of both the solid and surface modelling approaches.

Solid modelling developers are responding to the renewed demand for surface modelling by trying to add the ability to handle more complex shapes. Similarly, companies like Delcam are adding solids functionality to their surface modellers. These hybrid modellers, including Delcam's PowerSHAPE, aim to offer the speed and ease of use associated with solid modelling, together with the power and design flexibility of surfaces. It remains to be seen whether adding solids to a surface modeller or vice versa will give the most effective solution to the future users of rapid prototyping technology.

Fig. 1   LOM model of a gearbox housing for a Volkswagen Passat

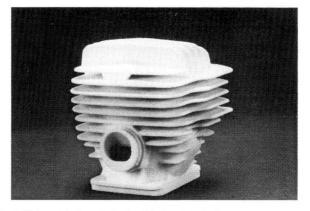

Fig. 2   SLS model of a cylinder head designed by Italian company Gilardoni

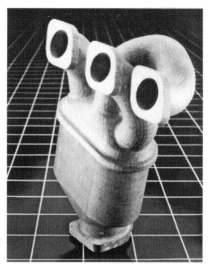

(a)

(b)

Fig. 3a and b Prototypes can be used as sacrificial patterns for casting models in the correct material

The introduction of CAD/CAM in the ceramics industry

Edith Groenewolt
Department of Industrial Design and Engineering, TNO Institute of Industrial Technology, Delft, the Netherlands
Jan Willem Gunnink
Department of Industrial Prototyping, TNO Institute of Industrial Technology, Delft, the Netherlands
Yvonne Vossen
Department of Ceramic Technology, TNO Institute of Applied Physics, Eindhoven, the Netherlands

SUMMARY

In many branches of industry, the computer has already made its appearance, in some cases a long time ago. In the ceramics industry, where there is great emphasis on the skill of craftsmen, that isn't yet the case. The EG-CRAFT project shows clearly that the introduction of computer aided design (CAD) and/or computer aided manufacturing (CAM) offers advantages to this industry too, and may even be absolutely necessary in order to satisfy the increasingly sophisticated demands of the modern consumer. This project is an initiative of the Dutch TNO and the University of Staffordshire. In Great Britain, a product developer (PCD), a pottery company (Denby Pottery) and a manufacturer of sanitary equipment (Shires) are involved. In the Netherlands, the participants are: MOSA Porselein BV, Koninklijke Tichelaar Makkum, Cor Unum BV, Plateelbakkerij Schoonhoven, Machinefabriek KIN, the Europees Keramisch Werkcentrum (EKWS), Koninklijke Sphinx and the Algemene Vereniging van de Nederlandse Aardewerkindustrie (AVA). The Finnish software company DeskArtes is also playing an active role by taking the comments and wishes expressed during the project with regard to the possibilities, and directly incorporating them into the adapted software.

In the project, the line of development from design to end product is worked through step by step: design, rapid prototyping, tooling and test run. By evaluating the whole process and its various components, and comparing them with the traditional process, it has become clear that one can reap benefits in various ways. For example, in the form of a better design, more design possibilities, a 3D model that can be shown to the customer for approval and prevents misunderstandings, time gained in the tooling process, better quality of the end product, and reduced costs. Not all the advantages apply to every part of every production process for every type of company in the ceramics industry. Sometimes the traditional method is faster or cheaper (for certain parts). Thanks to their familiarity with the possibilities and advantages of CAD/CAM, the ceramics companies can now weigh the various factors and choose the best (combination of) techniques.

1   THE BASIC PREMISE

Consumers are demanding more and more from products, and their desires change more and more frequently; the ceramics industry is no exception. Ceramic products have changed from utility items into fashion accessories. The life-cycle of ceramic products is becoming shorter all the time, and this increases the need to come up quickly with new or modified designs. On the other hand, in the ceramics industry the development and modelling phase, and the tooling too, are still carried out by hand. Consequently it is a long process and one cannot react quickly to new trends. In the past, the life-cycle of a ceramic form was five to ten years; now it is two to five years, and something new has to be introduced every year. True, the average development time has been reduced from two to four years to six to eighteen months, but that is still too long if a new product has to appear on the market every year. Moreover, development costs are high, especially in countries where wages are high. This means that one cannot react fast enough to market demand and that, frequently, not even the development costs can be covered.

For a good ceramic design, the following are necessary: knowledge of materials, knowledge of the market, a feel for design, and artistic qualities. The demands are high and so, therefore, is the price-tag of the design. An adaptation of an existing design should be quicker and less expensive. Practice shows that this is easier and faster with CAD than by hand, for the latter comes down in principle to completely modelling a form anew, because the existing model can only be modified to a limited extent. However, smaller companies lack the knowledge, the skills, the manpower and frequently the money to invest in the required technological development. Against this background came the initiative to set up the CRAFT project with the aim of reducing the time required for the development of ceramic products by using CAD, CAM and rapid prototyping. By introducing these new technologies, with guidance, to ceramics companies, they are eventually able to create more complex and varied forms and to react faster and more professionally to demands from the market.

1.1   The traditional development method
In the traditional way, a model is fashioned on the basis of hand-drawn sketches. The model, too, is shaped by hand, in clay for example. This model is modified however much as is required until the (external or internal) principle approves it. This same model is used as the master model for a plaster cast. The model is always scaled up by 20 per cent in connection with the shrinkage that occurs during production, with the disadvantage that assessment is always done on the basis of a model that isn't actual size.

Working from the plaster cast, a polyurethane mould is made, which is then used to produce a plaster production mould. Using this last mould, a small series of clay products is produced. These are fired in order to see how the product behaves during the firing process and whether the design needs modification. These modifications are made to the scale model, after which the entire process of the three moulds and the test run has to be done again. When the fired product satisfies the requirements and wishes, the required number of plaster moulds is fashioned from the polyurethane mould for the total production run.

1.2   The new product development method

The traditional method has the following disadvantages, sometimes with enormous consequences not only for the processing time and costs, but also for quality:

1. The freedom of form is limited because only rotationally symmetrical products can be fashioned.
2. It is a 'trial and error' process in order to find out whether the designed product can actually be produced in the intended form.
3. Assessment of the design is done on the basis of a scale model.

With the introduction of CAD, CAM and rapid prototyping and tooling, the new production process could look like Fig. 1.

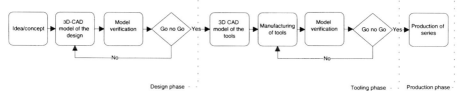

Fig. 1

In general terms, this new product development method offers the following advantages:

1. The designers are given an enormous amount of freedom in design by CAD's unlimited form possibilities in the idea and design phase.
2. Repetitive work does not have to be done by hand any more. With CAD, modifications can be made and evaluated quickly.
3. By means of CAD/CAM and prototyping techniques, models can be made quite quickly to true scale, which makes assessment easier and benefits quality.

In order to successfully introduce this new manner of working for the most common production techniques, it is especially important to ensure that the employees of the various participating companies can work independently with CAD/CAM in the end. For that reason the project was set up in stages as follows:

Phase in progress	Learning impetus	Content	End product
1. design	familiarity with software	• workshops • practice days • mastercourse	design for further realization
(2. decoration)	idem	workshop	
3. rapid prototyping	• data transfer • choice of technique • design assessment	• discussion • assessment sessions	a 3D CAD model
4. tooling	modelling of auxiliary tools	• discussion • assessment sessions	auxiliary tools
5. test run		assessment sessions	products
(6. decoration)			

During the project, three of the Dutch industrial partners developed a new ceramic product using the new product development method. The results were evaluated at the end of each phase. The results were incorporated in a manual that enables the companies to carry out the new method independently.

Company	Product	Production technique
Koninklijke Tichelaar Makkum (KTM)	Juice jug	RAM pressing
Cor Unum (CU)	Cup Saucer	Casting Forming
MOSA	Cup Saucer	Roller-head Plastic forming

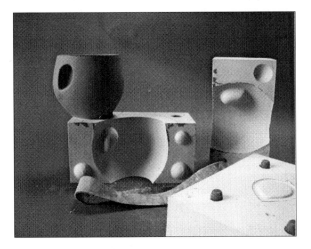

Fig. 2 Casting of the cup (Cor Unum)

2   THE DESIGN PHASE

There are numerous design programmes for various applications. Most are intended for the designing of technical products and are used by technically oriented people. Ceramics designers usually have an artistic training, and ceramic products usually have flowing lines. Prior to the CRAFT project the choice had already been made: the design program of the Finnish company DeskArtes was preferred because of its user friendliness and flexibility. In a series of workshops and a project of the TNO with MOSA, Koninklijke Tichelaar Makkum and the AVA, it had already been demonstrated that this design program was suitable in principle for use during the whole development process. The parties involved were agreed that a thorough training is an important success factor, because this gives sufficient opportunity for practice and the exchange of knowledge and experience, whereby problems can be recognized and solved.

For that reason, ample possibilities for this were built into the CRAFT project, in the form of workshops, practice days and a mastercourse. Moreover, after the workshops, the computers with the DeskArtes software were moved to the participating companies. Once the designers had some idea of how to use the program and they could actually move on to free design work on the computer, they became very enthusiastic about its many possibilities.

3   DECORATION

Some participating companies were very interested in putting decorations on curved surfaces. The question was whether the decoration transfer could be developed using CAD/CAM. Up to now, decorations are stamped out in the flat plane, which forms the basis for the transfer. However, it is impossible, in principle, to transfer a curved surface to a flat surface without deformation. Apart from wishing to do this without deformation, they also wanted to make modifications easily to existing decorations in order to use these for new models. Another wish was to manipulate the decoration in order to be able to make positives and negatives for the silkscreen printing technique. DeskArtes has made a list of these wishes and requirements and is working on them outside of the CRAFT project. This article therefore contains no more information on the matter.

4   RAPID PROTOTYPING

An object can be properly assessed on a screen only up to a certain point. The form, dimensions and functionality can be best discussed and assessed using a demonstration model. Using 3D CAD, and by means of rapid prototyping (RP), demonstration models can be made relatively quickly and simply. There are various RP techniques, distinguished in general terms according to the method of construction (by layers or by drops), the materials employed (paper, thermoplastics or wax), the accuracy and the associated costprice. TNO owns various RP machines. In order to make demonstration models with these, it is necessary to convert the geometrical information about the design into STL files (Structural Triangulation Language), in which the (curved) surfaces are described with triangles. The smaller the triangles, the greater the accuracy. Within the project, exercizes were performed in the conversion of design data into STL files and the transfer of data from the ceramics company to TNO. A fast and inexpensive method was chosen: Multi Jet Modelling (MJM). In this method, a model is built up layer by layer with the aid of a type of inkjet printer, except that the printer uses wax rather than ink. MJM models are excellently suitable for a (first) assessment of the form and dimensions (within the limits of the dimensions that can be produced with MJM: 25 x 20 x 20 cm). On the other hand, they are quite fragile, and in order to get a better impression of functionality (how the design sits in the hand, heat conductivity, leakage from spout), an FDM model (Fused Deposition Modelling) was made of a number of designs. In FDM, a semi-liquid thermoplastic material is deposited from an extrusion head in successive thin layers.

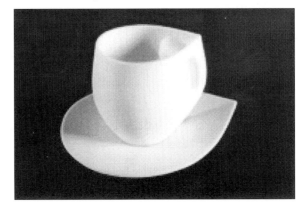

Fig. 3 MJM model of cup and saucer (Cor Unum design)

Fig. 4 MJM model of cup and saucer (MOSA design)

On the basis of the RP models, which were jointly discussed and assessed by all the participants, the following steps were taken:

- With regard to KTM's juice jug, attention was drawn to the ease of removal from the mould.
- The design of KTM's cup was modified because the handle was badly proportioned.
- The development of KTM's dish was called to a halt.
- The handle of CU's cup was not well formed, and because of this it becomes too hot.

- The saucer had to be modified because it was not possible to get a finger under the saucer in order to pick it up. Moreover, the proportions of the cup and saucer didn't seem to match each other.
- No comment was made on the cup and saucer of MOSA.

These deficiencies were not visible on the screen and would have negated the benefit of computer design if they had been discovered later in the process. The RP models thus doubly repaid the cost of the investment.

Com-pany	Product	Tech-nique	Material	Labour costs (NLG)	Equipment costs (NLG)	Process time (hours)
KTM	juice jug	MJM	wax	total 190	total 1300	total 25
	dish	MJM	wax			
	cup	MJM	wax			
MOSA	cup & saucer	MJM	wax	190	800	15
	cup & saucer	FDM	ABS	190	1210	30
CU	cup & saucer	MJM	wax	190	710	30
	cup & saucer	FDM	ABS	190	1550	30

5   TOOLING

The development of tools (master forms, dies, roller-heads, matrices, moulds etc.) forms an important part of the development process. The 3D CAD design of the (auxiliary) tools is made using the 3D CAD design of the product that is to be made. In order to decide how the (auxiliary) tools can best be made, one first has to answer such questions as:

- Which production techniques are used in the production of the end product (e.g. pressing, casting, roller-head)?
- What are the geometric demands this places on the tools (for example, requirements for attaching tool components)?
- How big is the run?
- What are the demands with respect to surface, form, and product quality?
- How many pieces are necessary for the construction of the plaster mould?
- What technique is used to make the tool components (milling, turning, casting, etc.)?

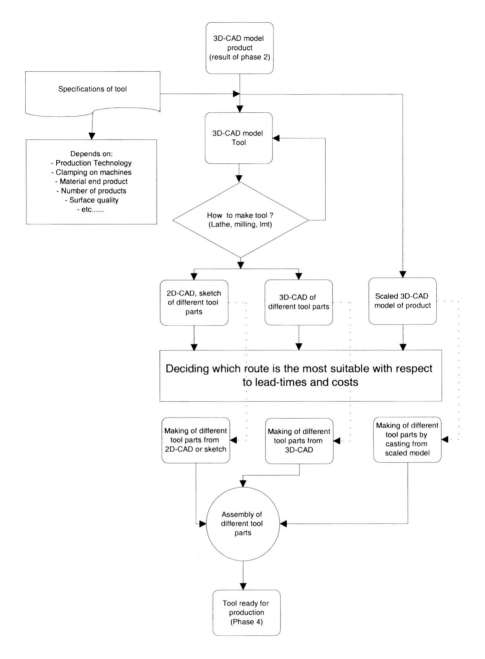

Fig. 5 Various tooling routes

Depending on the route chosen, the master form, its negative or the scaled-up model is modelled in 3D CAD and stored in IGES format. This makes the data suitable for processing via CAM. In the CRAFT project, the actual production of the tools was contracted out.

Fig. 6 Master form, plaster working form, and roller head for cup (MOSA)

6   TEST RUN

The (auxiliary) tools were used to make the plaster product moulds or working forms with which the eventual ceramic products were realized. The test runs showed clearly that both the tools and the end products satisfied the quality requirements. The products from the test run were also used to make a comparison with the traditional process with regard to the hours and costs in the various phases. Moreover, the quality of the designs, the prototypes, the (auxiliary) tools and the end products was evaluated.

For the sake of honesty, a number of remarks have to be made regarding the comparison between the new process and the traditional process:

- The new process was also a learning experience, which is to say that extra time was devoted to learning and practising. A retrospective estimate was made of the hours and costs for the CAD/CAM project without the tutorial aspect.
- The products developed using CAD/CAM were not developed along the lines of the traditional method. Experienced product developers and designers have made an estimate of the number of hours and the costs that the traditional method would have involved.
- CAD gives the designers more freedom than the traditional method of design. There is a high probability that these designs would never have been created using the traditional method.
- The repetitive procedures that form a large part of product development in the design of porcelain products are not included in this study.
- The elapsed time does include the workshops and the training programme, but not the production of plaster moulds.
- The MOSA design had already been tested at the start of the project.

Taking these remarks into account, the comparison between the new and the traditional development methods gives the following results:

Product	Time saving	Cost saving
KTM juice jug	On the basis of the current design, no direct reduction in process time is expected.	Circa 15%
CU cup and saucer	30 to 45%	20 to 30%
MOSA cup and saucer	Up to 40%	None, but if the design was done with CAD, a cost saving would be possible after all.

If the making of prototypes, models or auxiliary tools were to be subcontracted to companies specializing in CAM, this could render an internal time saving of 50 to 80 per cent. That time could be devoted to the development of new products.

7 NEW OPPORTUNITIES

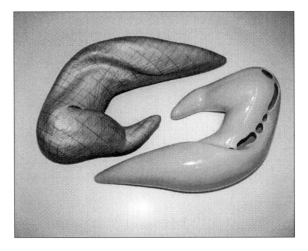

Fig. 7 This vase could never have been designed using traditional methods

Apart from the fact that working with the computer offers much more freedom to design, so that products are created that were never imaginable using the traditional method (see photo 5), much more is also possible in other parts of the process. During the CRAFT project all kinds of ideas about possible new applications and techniques came up. In order to make best use of the possibilities of CAD/CAM it is important that the people who work with it should, as far as possible, leave behind the traditional ways of thinking and working. One example. For some products, the master model is the positive of a product, because the standard casting technique demands it. With CAD/CAM, one is easily inclined to mill the same master form from aluminium or PVC. But with CAD/CAM it is just as easy to produce the negative of the master mould. A standard master mould can be made again using the usual casting techniques. In some cases, a negative requires less machining than a positive, and if the extra casting stage costs less than the extra milling for the positive then a net gain has been made. Another example of the advantages that can be gained is working with auxiliary tools of modular construction. The great accuracy with which the components can be defined and made using the computer guarantees that the components fit together well. Some more advantages:

- New products within a product line (for example, a set of plates in various sizes) can be designed much faster on the computer (than traditionally, on paper). New variants can be created quickly in CAD with a few simple clicks with the mouse.
- CAD can also be used well as a marketing tool: clients' wishes can be incorporated immediately in the product design on the screen. The modifications are visible on the spot and that is especially interesting as regards decoration.
- In the future the design program can be extended with calculating programs that forecast shrinkage and decorations during drying and firing. In this way, the number of repetitions in the design process can be reduced.

Quality
In addition to the gain in time and costs, the quality of tools and end products is also better with CAD/CAM. Among other things, this shows in:

- clean lines, sharp edges, and good surface qualities;
- better-fitting form components;
- improved product stackability;
- exact copies of any damaged master forms without loss of quality;
- optimization of design details is possible, which sometimes already leads to a reduction of the design layers.

8   IN CONCLUSION

Although the targeted savings in time and costs of 50 to 70 per cent have not (yet) been achieved, it is expected that these will be achieved when working with CAD/CAM becomes integrated as standard in the product development process. During the CRAFT project, the participants themselves discovered the many advantages and possibilities of working with CAD/CAM, which can only facilitate its introduction into daily practice. On the other hand, there are a number of pitfalls. It is important that people realize that errors in the CAD design are reproduced on a one-to-one basis by CAM. For that reason, it is important, by way of checking, to make a demonstration model of a (modified) design by means of rapid prototyping. Subcontracting production work (prototypes, models and auxiliary tools) helps to save time, but only with good planning and providing the various parties do not communicate only by means of computers. Apart from that, working with good toolmakers who have experience with all kinds of manufacturing industries can lead to interesting cross-fertilization. When the CRAFT project has finished, TNO will continue with the development of support activities for the Dutch fine ceramics industry in the field of product development and design technologies.

Product design and development – reverse engineering

Eur Ing Vinesh H Raja
Warwick Manufacturing Group, School of Engineering, University of Warwick, Coventry

ABSTRACT

At the heart of a company's competitiveness is its product development process. For many leaders in the automotive, aerospace, consumer products, and entertainment industries increased success means optimizing their product design and development process to achieve both increased productivity and efficiency, and the reduction of cost and time to market.

Reverse engineering is the process of duplicating an existing component or subassembly, without the aid of drawings, documentation, or computer model data, by using engineering analysis and measurement of existing parts to develop the technical data (physical and material characteristics) required for successful commercial reproduction. Simply put, it is doing whatever is necessary to reproduce something.

At present, 3D scanners are being used to capture 3D model data for reverse engineering of mechanical parts and organic shapes. This technology is not only used by the manufacturers but also by the medical industry for scanning bones to create prosthetics and implants and by video producers who capture difficult-to-create shapes for the television and movie industries.

This paper discusses novel applications of reverse engineering techniques for ceramics industries. The work undertaken is part of CERAM Sector Challenge Project funded by Department of Trade and Industry (DTI) in the UK. The case study focuses on use of reverse engineering techniques within the product development process for a tile manufacturer. This work has compressed H & R Johnson's product design and development time from 49 days to five days.

1   INTRODUCTION

Improving the efficiency of new product design and the introduction of new products to the market has become critically important, as product life cycles grow increasingly shorter. It has become widely accepted that the product development activity is the major driver of the downstream costs and quality. It is also recognized that reducing time to market requires the simultaneous engineering of products and processes, and getting into production with a minimum re-work and non-value added activities. Moreover, market trends, such as those depicted in Fig. 1, are set to continue and manufacturing companies need to take account of this. Shorter product life cycles combined with the need to provide a wider range of products more quickly means increasingly new designs to be incorporated, simulated, and analysed in the shortest possible time.

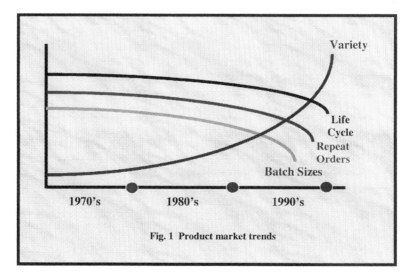

Fig. 1 Product market trends

In order to remain competitive in today's demanding markets, businesses are under increased pressure to deliver products in diminishing lead times, which meet the functional and aesthetic requirements of the customer at the right cost, i.e. faster, better, cheaper.

Advancements in CAD and CAM related technologies have helped to reduce product introduction costs and lead times, and improve the integrity of the information generated. However, at a number of stages within any product introduction programme, value-added work must be completed in isolation of the digital domain, as engineers will need to interact physically with their designs. From the point where the conceptual design leaves the drawing board to the stage where tools are manufactured for the production environment, electronic representations of a design will be shaped by feedback from functional and aesthetic tests completed on physical prototypes. The information generated during these tests is often held within the manually optimized features of the component. This information needs to be captured in an electronic form in order to support the computer-based downstream activities within the product introduction process.

The process of digitally capturing the physical entities of a component, referred to as Reverse Engineering (RE), is often defined by researchers with respect to their specific task (**1**). Abella described RE as 'the basic concept of producing a part based on an original or physical model without the use of an engineering drawing'. Yau, Haque, and Menq (1993) define RE as the 'process of retrieving new geometry from a manufactured part by digitization to modify an existing CAD database'. These concise definitions (amongst many) describe specific functions covered within the global term 'Reverse Engineering'. The recent growth in RE related technologies has helped to further broaden the scope of this valuable engineering process.

The range of RE functions possible – within the constraints of today's technology – can be categorized under three sub-headings.

- Surface generation (CAD)
- Tool-path generation (CAM)
- Feature analysis and inspection

These three RE options – shown in Fig. 2 – all occur downstream of the data capture process.

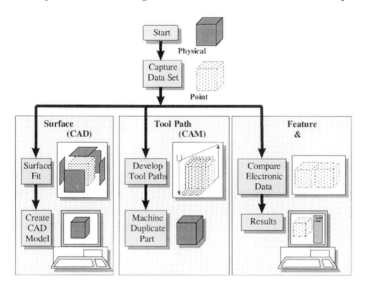

Fig. 2 The scope of reverse engineering

The first step of the RE process is to capture information describing the physical components features. 3D scanners are employed to 'scan' the component, producing clouds of points, which as a whole define the surface features. These scanning devices – which are available as dedicated tools or as additions to existing CNC machine tools – fall into two distinct categories: contact and non-contact (also referred to as tactile and remote sensing respectively).

1.1   Contact scanning devices

These devices employ contact probes which automatically follow the contours of a physical surface as depicted in Fig. 3. Within the current market place contact probe scanning devices, which are based upon CMM technologies, offer an accuracy which surpasses that achieved by non-contact techniques (approximately +/– 0.01–0.02 per point digitized). However, depending upon the size of component being scanned, contact methods can be slow as each point is generated sequentially at the probe tip. Tactile device probes must deflect in order to register a point, hence a degree of contact pressure is maintained during the scanning process (10–80 gms approximately). This contact pressure limits the use of contact devices, as soft, tactile materials – such as rubber – cannot be easily or accurately scanned.

Fig. 3 Contact scanning: touch probe[1]

1.2   Non-contact scanning devices

There are a variety of non-contact scanning technologies available on the market, all of which capture data without physically interacting with the component. Non-contact devices use lasers, optics, and CCD sensors to capture point data as depicted in Fig. 4. Although these devices capture large amounts of data in a relatively short space of time, there are a number of issues related to this scanning technology. The accuracy of non-contact scanning is still somewhat limited; tolerances as poor as +/– 0.5mm (for each point captured) are not uncommon (**2**).

Fig. 4 Optical scanning device[2]

---

[1] Courtesy of Renishaw plc.
[2] Courtesy of Newport Ltd.

Non-contact devices employ light within the data capture process. This creates problems when the light impinges on shiny surfaces, and hence some surfaces must be prepared with a temporary coating of fine powder prior to the scanning process.

1.3   H & R Johnson

H & R Johnson Tiles Limited is the UK's leading tile manufacturer producing wall, floor, and fireplace tiles as well as associated products. Their success is based upon a reputation for excellence in quality and design of products, which are supplied to both the consumer and contract markets. Innovation has enhanced their reputation and they were the first to develop and patent the 'universal edge' – a unique tile edge which allows even spacing of wall tiles, making their application simpler. Research, advertising, and distribution allow new products, created by leading British and Continental designers, to be presented to the public in the best possible way.

Exports make up an increasing share of the business. H & R Johnson products are sold in Europe, the Far East, Middle East, Australia, America, and South Africa. They have a thriving export business and take a pride in the fact that their products are held in high esteem throughout the world. The company pursues a policy of continuous improvement, which includes several recycling initiatives. In April 1997 they were awarded The Queen's Award for Environmental Achievement in recognition of their unique ceramic waste recycling project which saves over 5000 tonnes of waste being tipped in landfill sites each year.

As an innovative organization, H & R Johnson has done a quite a lot of work as part of the continuous improvement program. They felt the next step was to conduct a Business Process Re-engineering exercise within the product design and development department. They are looking at streamlining the product design and development process. Market research conducted by the company showed that increasingly consumers find tiles with natural organic surfaces far more aesthetically pleasing than human created patterns.

2   ORIGINAL PRODUCT DESIGN AND DEVELOPMENT PROCESS

Figure 5 depicts the original product design and development process. It starts when the marketing department conducts a detailed market research, leading to product specification and proposals. During this process, varieties of tile designs are considered and sketches of the new tile design are generated. From the product specification and proposals process, the model creator will sculpt the tile in clay from the sketches. This is a very labour intensive and time-consuming process. A resin tile is produced from the clay model, which is used as a master model for the production of a graphite electrode. The die cavity is produced by use of the spark erosion technique using the graphite electrode. The die produces the required set of tiles for design validation. Total elapsed time for the original product design and development process is 49 days.

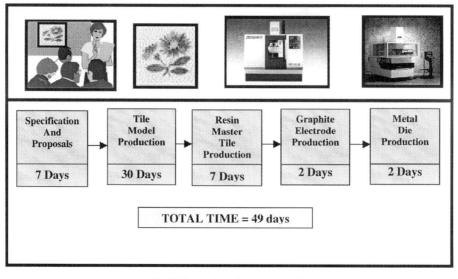

**Fig. 5 Original product design and
development process**

3   NEW PRODUCT DESIGN AND DEVELOPMENT PROCESS

The re-engineered product design and development process is as depicted in Figs 6 and 7 for
hard and soft surfaces respectively. The new product design and development process starts
with the market research leading to specification and proposals. This leads to a physical tile
designed by attaching a pattern required on to the surface of an existing tile. The physical tile
is then scanned using a standard paper scanner. The scanned image of the tile surface design
is exported to a package that allows the two-dimensional (2D) data to be modified by editing
'Z' values of the pixels into a 3D tile surface. This editing is a manual process, very time
consuming, and prone to errors.

The other alternative to standard paper scanners is 3D scanning systems. These systems can
be contact or non-contact. For hard surfaces both contact and non-contact is suitable, while
for soft surfaces only non-contact scanning is suitable. A 3D scanning system eliminates the
manual editing of pixels to create 3D data sets from a 2D scanned image. From the 3D data
set, a polygonized model is created. From the polygonized data, CNC programs are developed
and post processed for the machine tool to be used for machining the die. A die is produced
from resin or metal, depending on whether the tile design is at prototype stage or production
stage.

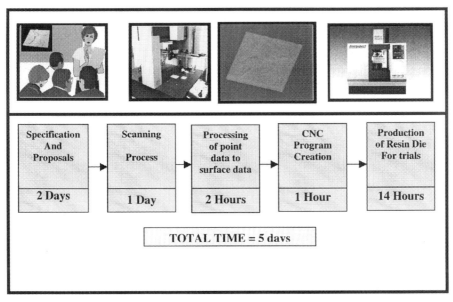

Fig. 6 New product design and development process for hard surfaces

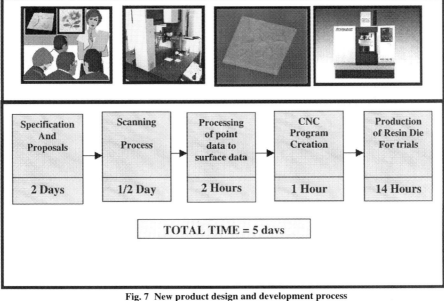

Fig. 7 New product design and development process
for soft surfaces

4 VALIDATION OF NEW PRODUCT DESIGN AND DEVELOPMENT PROCESS

It was felt that the best validation is for the company to apply the new process for a potential project. Figure 8 shows the tile design, which is created by gluing a set of leaves on to an existing tile blank. The tile is scanned using a non-contact scanner and a point cloud data set is created. The point cloud is polygonized and CNC tool paths are generated. The post processed CNC program is executed on the machine tool and a soft tool for the new tile is produced. The prototype tool is used to press a set of tiles and the tiles are eventually lacquered.

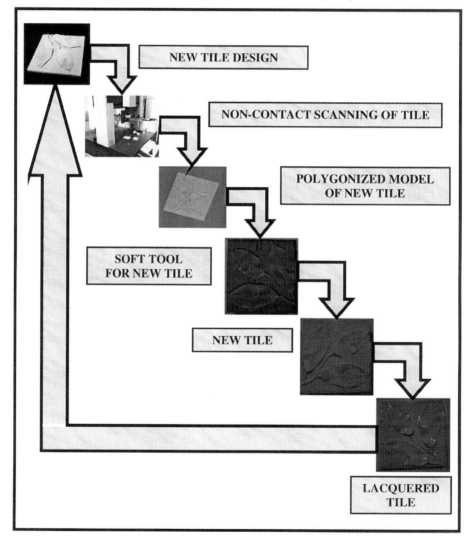

NEW TILE DESIGN

NON-CONTACT SCANNING OF TILE

POLYGONIZED MODEL OF NEW TILE

SOFT TOOL FOR NEW TILE

NEW TILE

LACQUERED TILE

Fig. 8 Closing the loop between production and design

5   DISCUSSION AND CONCLUSIONS

The new product design and development process compresses the tile development time from 49 days to 5 days. The new process requires H & R Johnson either to make investment for the purchase of scanning technology or to use scanning technology on a bureau basis. Over a period of time, spark erosion technology will not be required, hence there will be saving on maintenance costs as well as any future investments in acquiring a new machine. This process also gives H & R Johnson the agility required for a quick response to customer taste, thus make them more competitive.

Since all the data related to tile design and manufacture can be in an electronic form, H & R Johnson can quite easily provide a web based virtual tile collection. This would provide their prospective customers a web-based system to evaluate different permutations before ordering tiles through a web-based tile shopping mall.

The author wish to acknowledge the Warwick Manufacturing Group and their partners, H & R Johnson and the DTI for providing the resources, validation of the new process, and providing funding, respectively.

REFERENCES

1   **Motavali, S.** and **Shamsaasef, R.** (1996) Object-oriented modelling of a feature based reverse engineering system. *Int. J. Computer Integrated Manufacturing*, **9** (5), 354–368.

2   **Yau, H. T., Haque, S.,** and **Menq, C. H.** Reverse engineering in the design of engine intake and exhaust ports. *Manufacturing Science and Engineering*, (ASME).

3   **Hsieh, Y. C., Drake, S. H.,** and **Riesenfeld, R. F.** (1993) Reconstruction of sculptured surfaces using coordinate measuring machines. *Advances in Design Automation*, **2**, 35–46, (ASME).

4   **Galt, A.** (1997) Scanning technology – the missing link. *Prototyping Technol. Int.*, 121–124. Also in *Manufacturing Science and Engineering*, 1993, **64**, 139–148, (ASME).

Reducing the risk of new product development by utilizing rapid prototyping technologies

Steve May-Russell and Martin Smith
Smallfry Industrial Design, UK

1 OVERVIEW

The aim of this paper is to demonstrate the application of rapid prototyping (RP) technology as a compelling tool that can provide benefits throughout the process of developing new products. It is based on Smallfry's experience of product development programmes across a wide range of product types for clients around the world. It aims to provide an insight into the way in which RP technologies have been integrated into the strategic design development process that Smallfry employ to ensure that the products under development are marketable, manufacturable and ultimately profitable. In particular the focus is on how RP can be used to demonstrate and test the many attributes of a new product, as it is developed, that fall outside the direct domain of product engineering.

The main conclusions to be drawn are that RP can provide major benefits throughout any organization involved in the development of product hardware, from the Sales and Marketing team through to Quality Control and Finance. Specifically, there are serious benefits that RP can bring in the areas of market research, sales support, promotional material, and the ever-important product launch.

Utilized in the right way RP can also become a powerful communications tool to ensure that everyone involved in the development process fully understands and appreciates the product being developed. This can substantially reduce the inevitable risks involved in any product development programme by allowing all the critical decisions to be made with fully informed and equally based judgement, removing any misunderstandings and highlighting any potential conflict of interest between departments.

2   THE NEED TO GET THE PRODUCT RIGHT

The whole process of delivering a new product offering is fraught with danger and the risks involved are substantial. New product development (NPD) seems to be widely acknowledged as a necessary evil. The drive behind NPD projects is, all too often, a fall in the sales of an existing product line or a new player in the market offering a better proposition. From this position of 'catch up' the investment in a new product needs to be well spent to ensure survival. Moreover there is the overriding certainty that if a company continues blindly to do what has brought them success in the past, it will inevitably lead to their downfall – there will always be a competitor eager to steal the market. The smart company will leapfrog the competition and constantly strive to develop fresh, new products both to excite the existing market and create new opportunities. This is not without its risks, but fresh product ideas are vital to guarantee continued success and prosperity, and the long-term rewards that can be gained are obviously far greater.

To compound this need to produce fresh product ideas, it is important to appreciate that the product that is chosen for development is appropriate for its market and offers genuine benefits that are tangible to the intended customer. A product that reaches the market that has been ill conceived or falls short of the market's expectations can cause irreparable damage. Other than the obvious waste of the investment and resources used to bring the product to life, there is the damage that a poor product can do to the existing product lines that it was intended to compliment or supersede. More importantly still, there is the potential for a poor product to seriously devastate the brand it represents; it is frightening how often the product itself is overlooked as the most direct embodiment of what the brand stands for. A product that is uncomfortable or awkward to use, or that consistently fails, breaks down, or merely frustrates the user, will still carry the brand name as a constant reminder of the company responsible throughout its life. Therefore every opportunity should be taken to ensure that the product being developed is realized in the right way. This is where RP technology can provide a fast, reliable way of assessing the design to find any shortcomings sooner rather than later.

3   RP TECHNOLOGY AS A STRATEGIC TOOL

The process of delivering commercially successful new product offerings depends on creating the right product to meet customers' evolving requirements *and* delivering them through to market in a way that supports the business objectives of the company behind it. The strategic approach to NPD that Smallfry have pioneered is founded on ensuring that all the factors that will contribute to the success or failure of a project are determined, considered, brought together, and resolved in the most appropriate way. This not only encompasses the manufacturability of the product, but also its functionality, financial viability and, most importantly, its marketability. A crucial factor in this strategy lies in ensuring that the design intent behind a new product offering is clearly communicated to all the parties involved in the project at various key stages of development. RP technology provides a variety of realistic and distinct methods of demonstrating the design as it develops, allowing the team to check that the product proposition remains on target in fulfilling all of the criteria that will contribute to the final success of the product.

Fig. 1 CNC model used to validate the size and external form of a new cycle light. The model was also used for an early exhibition launch and for consumer focus groups, establishing the market while the design was still being developed. 'Shark' cycle light for Basta France SA

Within the context of the consumer product development programme, RP technology can take many forms, each providing its own particular benefits and limitations. Some of these prototyping techniques would not necessarily be recognized within the traditional framework of what is considered as rapid prototyping technology. For the purpose of clarity our working definition of what we regard as rapid prototyping technology in the context of this paper is the combined use of computer generated data and technology-based modelling to produce a first off that can be tested to identify any necessary changes. This definition encompasses the typical three-dimensional RP techniques such as SLA, SLS, MJM, LOM, and CNC. It also includes the down-stream reproduction of RP parts to produce short run batches and to replicate specific materials in both plastics and metals through rapid tooling technologies. However, modern computer technology also offers other techniques that provide a practical means of testing a product as it develops. These digitally based techniques include photo-realistic visualization, interactive software simulations, digital product animation, and virtual reality models. Utilizing all of these tools together provides the product development team with a vital means of testing practically every aspect of a new product that will contribute to its success *before* it finally reaches manufacture.

4   COMMUNICATION TO ENSURE COMMITMENT

Perhaps one of the most surprising roles of RP in terms of the overall business strategy lies in its ability unambiguously to communicate a design throughout an organization. NPD is a core business activity that needs to be managed properly and have the full support of all members of middle and senior management. To achieve the best results an NPD project will capitalize on the combined efforts of a well-integrated, cross-functional team that spans all levels of the organization. This team might comprise representatives from finance, marketing, engineering, production, distribution, and maintenance as well as customer groups and consumers.

Inevitably each will see things from their own perspective and naturally pursue what best satisfies their own objectives, budgets, and resources. Most will not have a background in design and development and should not be expected to be conversant with the language of product engineering. RP models help to bridge the gap for non-technical people by making incomprehensible engineering specifications, or intangible, nebulous concepts generated by the creative team, into tangible objects that everyone can understand.

Solid RP models, even from early stages of the design process, can quickly and convincingly show the external design shape, overall sizes, aesthetic effects, and ergonomic aspects of a product in a way that is much more readily comprehensible than any engineering scheme or designer's sketch. By literally putting product ideas into everybody's hands RP removes all preconceptions about the intended product and ensures that the team can discuss the product on the same footing. Equally photo-realistic images, animations, and interactive on-screen presentations can help to demonstrate convincingly the objective of the new offering in terms of its appearance, functionality, and features. This allows the team to assess the benefits that the design is intended to offer and whether they are being offered in the right way. The more clearly these aspects can be demonstrated, the better the chance of avoiding the ominous reaction to the first tooled product; 'it's much bigger/squarer/more difficult to use than I imagined'.

Fig. 2 Exhibition literature produced entirely from computer visualization. This demonstrated the benefits of the product at a key exhibition to provide maximum exposure for the next season's range. 'Shark' cycle light for Basta France SA

Beyond preventing potential problems and delays brought about by misunderstandings, RP, if used imaginatively, can produce stunning results outside the direct development team by exciting and motivating *everyone* involved in creating, approving, and delivering the new product offering. Gaining early commitment to an NPD project throughout an organization can be essential to maintaining the thrust behind a project and easing the progress through manufacture. RP provides material that can be used to demonstrate new designs, in a controlled manner, to personnel not directly involved in the project. This helps to promote a sense of ownership and avoids the 'not invented here' syndrome, generating and maintaining enthusiasm for the project and keeping everyone focused on a common goal.

5   APPROVALS AND VENTURE CAPITAL

It is the nature of NPD that time is an all-important competitive tool. An undisputed advantage of RP is that it can speed the design development process. However, it is perhaps less commonly understood that RP can equally speed the process of getting a new product through the red tape, approvals, and 'gate' targets that many large organizations require.

The accurate representations that can be produced by RP provide an ideal means of shortcutting lengthy approvals procedures. Rapid tooling techniques can produce short runs of components, in actual materials, for various accelerated life tests. These can range from environmental, IP, and EMC approvals to load tests, drop tests, and Beta site tests for electronics and safety equipment. While not providing an absolute substitute for testing actual product samples in every case, RP components can go a long way towards ensuring that the final product will not encounter unexpected problems in achieving its necessary approvals and ratings. Within the bureaucracy of large organizations, Smallfry have also used RP models as a means of accelerating new products through quality approvals and financial sign-off, rather than the usually expected first-off tooled samples. This has taken several weeks out of the overall time to market.

Outside the large organizations for which NPD is a regular activity, RP can be employed as a means of generating interest in a new product venture. Convincing photo-realistic images and hardware can again communicate the benefits of a new product proposal to potential backers completely outside the domain of product development. Short interactive presentations running on a computer screen have been successfully used to demonstrate the potential provided by new equipment in terms of physical capabilities and/or the user interface or product software. This allowed the companies in question to secure financial backing to develop a product idea through to the market that might not otherwise have seen the light of day. In other instances we have helped our clients to secure large contracts to develop new products for key customers based on a computer-based demonstration of the product's potential.

6   EASING THE MANUFACTURING PROCESS

The use of RP techniques to verify CAD data prior to the commissioning of manufacturing tooling is, perhaps, where the technology found its initial strength. As such the advantages in this area are widely documented. Suffice to say that the benefit of being able to put an accurate representation of a new component in the hands of those directly involved in its future manufacture, provides instant and clear understanding. This allows immediate feedback on the overall viability and, in many cases, this has shortcut weeks of poring over drawings or onscreen CAD models before identifying a potential problem.

Perhaps less widely considered is the advantage that RP components can provide in smoothing the downstream processes between moulding and the market. This includes the creation of the jigs and fixtures for printing, welding, and assembly of components. A full set of RP components also allows the sequence of assembly and resource requirements to be accurately evaluated to gauge the costs and timings necessary for full-scale production. This can help significantly with the planning and management of the production facilities. Automated assembly robots can be programmed and manual assembly workers can be familiarized with the products they will soon be responsible for. Again this can bring detail changes to light that may simplify and reduce costs of assembly before the component tooling is finalized.

By bringing forward the point at which all of these manufacturing processes can be started, any teething problems of manufacture can be identified and resolved early on, all the time speeding up the time to market and reducing the need for initial modifications.

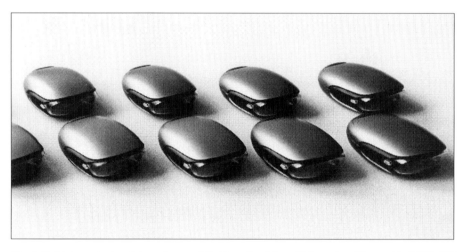

Fig. 3 Multiple product prototypes for a cycle light produced from SLA masters. These were used to arm the international sales team with samples, allowing them to fill their order books while the product was still being tooled. 'Shark' cycle light for Basta France SA

7   RP AS A MARKETING TOOL

Creating anything new is an evolutionary process. 'Right first time' is a fallacy within the context of NPD. Having identified an opportunity in the market, a typical elapsed time to deliver a new injection-moulded product is 9 to 12 months (obviously, this will vary depending on the exact nature of the product). It should not be assumed that the product that was wanted when the programme started is what will be wanted when the product becomes available. The consumer product market is dynamic and volatile. In order to guarantee the product meets the needs of the marketplace it should be repeatedly re-evaluated and verified to ensure it remains correctly focused. Perhaps the best advantages of RP to date are being seen by the sales and marketing teams both in the process of developing a product with the right features, and in supporting the product through its launch on to the market.

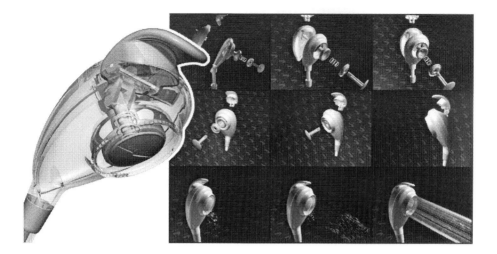

Fig. 4 Computer visualization and stills from a computer animation used to demonstrate the construction and function of a new product concept. The animation was used to sell the concept into key retail outlets to secure the essential routes into the market. 'Freeway' Shampoo Spray for Croydex (to view this figure in colour, go to colour section)

As a product is under development the obvious way to test its appeal and feature set is by gauging the reaction of key target consumer groups through market research. Again RP can be used very effectively to assist in this activity by providing material to test the market potential of a new product through consumer hall tests and user trials. Where this involves members of the public assessing a product proposition against a selection of real, existing products, they cannot be expected to make concessions for the product samples being short of real in their appearance, feel, and functionality. By providing small batches of product that are, to all intents and purposes, 'real', RP brings the most certain indication of the expectations of the market that could be wished for. As with the development team all ambiguity is removed and the sample users can provide genuine reaction to the product, as if they were assessing it at the point of purchase.

In other instances it is preferable that market testing provides only a limited exposure to the product, either for reasons of confidentiality, or to test only specific aspects of the product proposition. This is where animated presentations can be used to great effect. Moving images have far more impact than a still photograph. Technology allows the product to be brought to life in a real context and fly-throughs, moving parts, methods of assembly, displays, and modes of operation can all be demonstrated. The way that specific features are intended to function can be shown in a clear and compelling way, and the advantages they bring to the user can be highlighted. The overall effect is exactly like a TV advertisement selling the product's features and advantages.

Having utilized RP to ensure that the product under development has the right qualities to appeal to its intended market, the Marketing Department will also need a substantial amount of supporting material to enable its timely launch on to the market. All too often the overall time scale of the NPD programme does not allow sufficient time for the product packaging, as product samples are needed for establishing the overall sizes and shapes and the most suitable methods of packaging, as well as for commissioning the pack photography. This can substantially delay the launch date or distribution of the product to its intended outlets, missing valuable early sales opportunities.

RP prototypes and computer visualization have both become a boon to finalizing the packaging and distribution needs of new products. Initial 3D product samples are obviously of great help to the packaging designers, particularly where a product demands complex, structural packaging. Photo-realistic imagery can also provide major advantages in progressing the product pack, operating instructions, and support material. Smallfry have utilized computer visualization to produce all the product images needed for entire ranges of packaging. Not only are these images indistinguishable from real photography, it is also possible to achieve details, cutaways, and see-through views that the real camera would never allow. Equally all the traditional problems of photography, such as unwanted reflections, dust, shadows, or wires and supports, all disappear in the world of computer image generation. If all of these factors are carefully managed RP allows the product packs to be developed ready to coincide with the first product off the production line.

Prior to a new product launch, generating the necessary market buzz can be essential to the initial success of a product. This starts with pre-selling the product to the sales team to allow them to understand the new products they will have available for the coming season. As we have already discussed, RP has a vital role in communicating new designs to everyone involved. Once the sales team has been introduced to the product, RP allows them to be armed with sales samples a vital two or three months before the actual product can be shipped. These RP models can prove indispensable for negotiating with potential customers by demonstrating exactly what they can expect from a new offering. Customers won't buy what they haven't seen, so being able to show a customer a forthcoming product strengthens the sales team's negotiating leverage. This can guarantee inroads into the essential retail outlets for the next 2 or 3 years sales.

Equally, providing working product samples for exhibition at trade shows and seminars, where there are essential launch windows that have to be hit, is another area where RP can help to shave valuable weeks off the apparent time to market. It should be noted that the design of the product does not have to be complete to allow the pre-selling process to begin. The right combination of RP tools can produce a persuasive package to explain all the features and functionality of the final proposition at the earlier stages of the development programme. Convincing presales literature can be created as soon as the external package has been generated on the CAD system. To the outside world this can appear to show a fully finalized product. Again this can help to ensure retail space in key outlets is reserved for your forthcoming products for the next season.

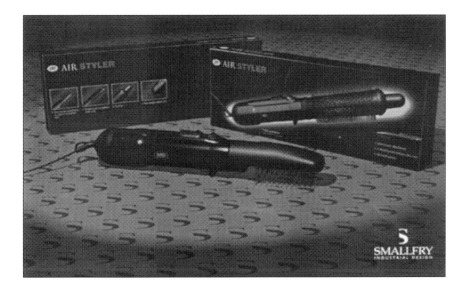

Fig. 5 Smallfry used computer visualization to generate all the packaging images for a range of hair care products. This allowed the packs to be completed ready for the first product off the production line. This computer generated image showing the product and pack together was used for departmental approvals. Hair care range for Boots the Chemists

8   CONCLUSION

The wealth of RP processes now available has brought nothing short of a revolution in the process of new product development. From a designer's perspective, the combination of computer aided design and technology-based modelling techniques has spurned advantages' beyond their wildest dreams. The peace of mind that comes from having an RP model that bears out the designer's intentions, and validates its function and manufacturability, is immeasurable. There is now no excuse for not delivering exactly what the RP has confirmed is possible.

In the wider context of an overall NPD strategy, RP can help to reduce substantially the inevitable risks in the route from product concept to commercial success. It is vital to be certain that the product you are developing embodies the best resolution of all the criteria that will affect its potential for success out in the market. RP makes intangible designs tangible to everyone with a vested interest in the product, both within the organization and outwardly, through the chain of supply, to potential end users. Each of these people is a 'customer' with their own criteria and opinions. The sooner this wide range of needs can be taken on board and tested, the less time and money is wasted on developing features or ideas that do not fit in with the needs of the business plan and the desires of the intended market. RP provides the necessary material to thoroughly prove-out the design and, if applied intelligently, help with virtually every aspect of product development including finance, planning, manufacture, research, testing and, especially, sales and marketing. At each stage RP provides a full, clear appreciation of what is being committed to, reducing the risk of NPD.

Stress analysis using rapid prototyping techniques

G C Calvert
Rover Group Ltd, Experimental Analysis, University of Warwick, UK

ABSTRACT

The Experimental Analysis Department within Rover Group, based at the University of Warwick, has been developing the use of experimental stress analysis techniques on models produced by rapid prototyping (RP) technologies. The techniques have been adapted for use with RP models include thermoelastic stress analysis (TSA) and three-dimensional photoelastic stress analysis (PSA). Work over the last two years has proven the validity of the techniques and their ability to provide full field stress analysis data from models generated directly from computer aided design (CAD) files.

This paper will present the business reasons for taking this approach and the advantages gained in shortening the product development cycle, the importance of validating a design prior to expensive working prototype build, and the cost saving benefits to be gained. A case study is presented on a gearbox splined shaft, using both thermoelastic and photoelastic analysis techniques, to demonstrate the technology.

1   BACKGROUND

There is increasing pressure within the automotive industry to introduce new models into the market place in a much shorter time frame, and at more competitive prices than ever before. To achieve this, product design cycles have to be shortened and costs substantially reduced. This means less design iterations or levels and therefore a reduction in the number of working prototypes. Along with this there is still a need to increase quality which means a greater requirement for confidence in the initial concept design.

Historically, experimental analysis techniques have had to be carried out at the 'working prototype' stage in the design/development cycle of a product. This is because, in the case of TSA, an actual part had to be available in order physically to subject the component to cyclic loading and hence obtain a surface stress solution. For PSA, it was possible to instigate the work earlier but invariably, due to the need for a physical article from which to cast a model in a suitable photoelastic material, the process took as long as producing a real working prototype. The techniques, therefore, were often too late to make radical changes to the design, as the cost of making late changes to tooling etc., in order to make the design more efficient, would have been prohibitive at this late stage. This often resulted in a compromised design and what was clearly needed was a method for early validation and correlation with finite element analysis (FEA).

One way of providing additional information is to utilize experimental analysis techniques by applying them to RP models which can be constructed direct from CAD data at the concept stage of design (1,2). By carrying out this combination of computational and experimental analysis at the concept stage of the design cycle, early confidence in the integrity of the design can be achieved without going through the usual design, build, and test cycle. This results in substantial cost and time savings, because the number of expensive working prototypes and the amount of testing time needed to validate the product have been greatly reduced. Figure 1 demonstrates this, with a comparison of the two methods. The old method showing low cost initially and then escalating costs, due to the build and test of real working prototypes, through to the end of the programme. In contrast, the rapid method shows an initially higher cost due to the early production of RP models and experimental testing alongside FEA work, but lowering costs towards the end of the programme due to savings in production tooling and testing of real working prototypes.

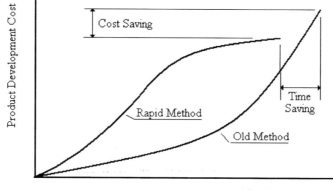

Fig. 1

2 THERMOELASTIC STRESS ANALYSIS

When a component is loaded there is a temperature change associated with the molecular displacement. When under compression there is a heat increase and when under tension there is a heat decrease (3).

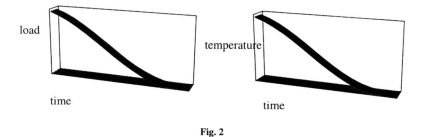

Fig. 2

This temperature change under loading is proportional to the sum of the principal stresses on the surface on the component (see Fig. 2).

$$(\sigma_1 + \sigma_2) = \frac{\Delta T . \rho . C_\sigma}{\alpha . T}$$

As these temperature changes are in the order of thousandths of a Kelvin they are not apparent in everyday life and also would be very difficult to measure. To solve this problem the component is cyclically loaded, effectively loading the component, unloading back to its starting load, loading, unloading, and so on. In this way the static load on the component is being modelled dynamically (Fig. 3).

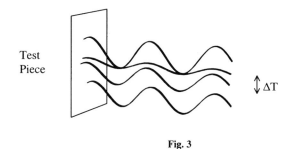

Fig. 3

By using a specialist thermal camera to measure the minute temperature fluctuations, it is possible to obtain a full-field stress map of a component.

The magnitude of the temperature changes is equal to the sum of the principal stresses multiplied by a constant for that particular material. The constant is calculated by measuring the thermoelastic response of the material on a system of known stress, for example a tensile test specimen or a Brazilian disc. Alternatively, if three properties – density, expansion coefficient, and specific heat – are known for the material, this constant can be calculated mathematically.

Thermoelasticity will provide a full-field stress map of a component made from most materials, including rapid prototyping resins. The boundary conditions will be accurate, and as long as the material remains elastic, it can be used even in non-linear problems. It does, however, require the component to be subjected to a cyclic load, and also optical access to the areas of interest. It should also be observed that the output is in the form ($\sigma_1 + \sigma_2$), the sum of the principal stresses on the surface of the component.

3   PHOTOELASTIC STRESS ANALYSIS

Photoelastic stress analysis, utilizes an effect known as 'birefringence'. When a 'birefringent' material is stressed and observed under polarized light, interference fringes are generated which coincide with principal stresses (4).

The technique can be used to obtain extremely detailed stress data throughout the component by carrying out three-dimensional photoelasticity. This involves constructing a detailed, full-size or scale model of the part under investigation from a birefringent material; in the case studied here, an epoxy resin from the 3D Systems Stereolithography RP machine. The model is then loaded in a representative manner, with scaled down loads and subjected to a 'stress freezing' cycle. This involves heating the model up to the material's glass transition temperature, at which point the Young's modulus changes and the model deforms under the applied loads. It is then slowly cooled, avoiding any uneven temperature, which could result in unwanted thermal stresses. During the cooling cycle the deformations and stresses are locked into the model. When viewed under polarized light, the three-dimensional model is a jumble of interference fringes, so in order to determine both magnitude and direction of principal stresses at any point, a slice is removed and observed under polarized light. Obviously care must be undertaken when removing the slice, to avoid exceeding the critical temperature of the material, otherwise the stresses locked within the model could be removed or added to, with disastrous results to the stress figures. The stress in the model can be calculated by knowing the 'fringe coefficient' of the material, and converted to actual component stresses by means of proportionality, between the model and component material, loading and dimensional parameters as shown below (5):

$$\sigma_1 - \sigma_2 = n \, f \, t \qquad \text{where:} \qquad$$

σ_1 = maximum principal stress
σ_2 = minimum principal stress
n = measured fringe order
f = material fringe coefficient
    (determined by calibration)
t = thickness of slice

To obtain component stresses:

$$\sigma_{Actual} = \sigma_{Model} \% \frac{P_{Actual}}{P_{Model}} \% \frac{L_{Model}}{L_{Actual}}$$

where: σ_{Actual} = actual stress
σ_{Model} = model stress
P_{Actual} = actual load
P_{Model} = model load
L_{Model} = model dimensions
L_{Actual} = actual dimensions

Photoelastic analysis is an extremely powerful and useful tool for the designer, due to its ability to provide full-field stress analysis both externally and internally. The problem in the past has been the long lead times needed to construct the models and in the analysis and interpretation of the results. Technologies such as rapid prototyping, automatic fringe analysis systems have removed these shortcomings, giving the technique a new lease of life as a cost- and time-effective experimental analysis tool.

4   CASE STUDY - GEARBOX SPLINED SHAFT

4.1   The component

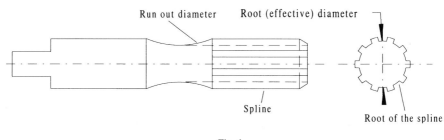

Run out diameter   Root (effective) diameter

Spline

Root of the spline

Fig. 4

Figure 4 shows the component under investigation, the objective being to determine the stress distribution over the length of the splines when subjected to a torsional loading. For confidentiality reasons, the stress values will not be published, but a comparison will be made between the thermoelastic and photoelastic analysis techniques.

4.2   Thermoelastic analysis

The thermoelastic analysis was carried out by constructing a twice full size RP model of the splined shaft and a female splined drive boss, through which to react the applied torsional load (6). The model was constructed on a 3D Systems Inc. SLA 500 machine, with a commercially available Ciba-Geigy epoxy resin. A loading rig was constructed, consisting of a lever arm mechanism attached to a single electrohydraulically controlled actuator, which allowed the required cyclic loading of the model. Cyclic loading in the form of a sinusoidal input was required in order to obtain the thermoelastic data, which were gathered using a CEDIP (7) infra-red array camera with appropriate software to display the output in terms of the summation of principal stresses as explained in Section 2. The CEDIP system allows easy focusing on to small detailed areas of the spline and has the ability to provide a reasonable stress solution within 30–60 seconds. The output is displayed below in Fig. 5.

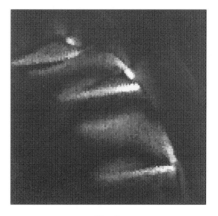

Fig. 5

A line scan was taken across the image to obtain a graphical display of the stress distribution along the spline roots and hence the maximum recorded stress level. Calibration of the material to determine the thermoelastic constant was carried out using a Brazillion disc, loaded in compression, which provides a known mathematical solution for stress in the centre, which can then be compared with the thermoelastic signal at that point. The resulting stress value for a real component can then be determined using similar proportionality rules to the photoelastic analysis as described in Section 3.

4.3 Photoelastic analysis

The photoelastic model was generated directly from the CAD data by means of the same RP machine, which was used for the thermoelastic analysis. In this case, once the model was assembled into the female boss, a static scaled down load was applied to simulate the maximum torque being driven through the splines and the whole loading rig was placed inside an oven for the stress freezing cycle. Over a 30 hour period the deformation and stresses were 'frozen' into the model spline as described in Section 3. The model was then removed and a series of slices taken at right angles to the length of the spline. The fringes generated can be seen in Fig. 6 for two of the slices, and from these photoelastic fringes, a picture of stress distribution along the spline's length could be determined. Unlike with the thermoelastic analysis, individual maximum principal stresses can be calculated with the photoelastic technique using the 'oblique incidence' method (8), which uses the principles of Mohrs circle. The results are therefore more detailed and precise stress concentrations can be determined.

Fig. 6

4.4   Comparison of results

The maximum stresses in the photoelastic analysis were found to be approximately 20 per cent lower than those found with the thermoelastic analysis, although general stress levels correlated well. This could be due to the positioning of the slices taken for analysis, which may have just missed the maximum area. Also it is very difficult to simulate pure torsion and certainly in the case of the thermoelastic analysis some bending of the shaft was inevitable due to the orientation of the loading arm which may have raised the stresses. Overall both the photoelastic and thermoelastic analyses showed that the stress distribution over the splined teeth was not uniform and that large variations in stress could occur due to manufacturing tolerances. The results highlighted the need for caution when predicting stresses on splines as the parallel finite element analysis assumed uniform loading on all spline teeth. The experimental methods both proved that assumption invalid.

5   CONCLUSIONS

The case study proved that rapid prototyping could be used to construct detailed models for experimental stress analysis at an early stage in the design cycle. The thermoelastic technique can be used to gather results extremely quickly, to provide correlation and validation with finite element methods of stress analysis. The photoelastic analysis technique obtained much more detailed information than the thermoelastic analysis but took almost three times as long due to the stress freezing cycle, slicing process, and more complicated analysis of the results. In order to become more effective, work is required on reducing the loading and analysis time. This is being addressed with several automatic fringe analysis systems becoming available, but still the loading and slicing stage is the major issue. Work is being undertaken at the University of Warwick to eliminate the slicing process and provide detailed stress analysis results with very low fringe orders, thus simplifying the whole process.

The choice of technique depends upon the complexity of the problem. However, there is no doubt that the advent of rapid prototyping, enabling almost instant models of components to be generated, has provided the means for early validation and enhancement of finite element analysis and the chance to reduce dramatically both costs and time in a products development cycle.

ACKNOWLEDGEMENTS

The author would like to acknowledge the following people for their help and assistance in this work: Richard Winder, Rover Group, Claus Schley, University of Warwick, Juergan Wolf, University of Stuttgart, and Peter Sobola, University of Warwick.

REFERENCES

1   **Calvert, G.** (1995) Rapid prototyping and its use for experimental analysis. The Engineering Integrity Society Annual Conference *Computers in Engineering a Boon or a Burden?*, April 1995, Sheffield Hallam University, UK.

2   **Driver, C., Winder, R., Calvert, G., McDonald, J., Bryanston-Cross, P. J.,** and **Udrea, D. D.** (1998) Rapid experimental conceptual analysis. Rapid Prototyping and Manufacturing 1998 Conference and Exposition, 19–21 May, Dearbourne, Michigan, USA.

3   **Winder, R.** (1998) Thermoelastic analysis of rapid prototype models. The British Society of Strain Measurement *Rapid Prototyping and Experimental Analysis Seminar*, 24 June, University of Warwick, UK.

4   **Sharples, K.** (1981) *Short Course Notes* (Photomechanics Ltd).

5   **Calvert, G.** (1992) Stress analysis techniques for composite materials. MSc thesis, University of Warwick, UK.

6   **Sobola, P.** (1999) Thermoelastic stress analysis of a gearbox spline. Third year BSc project, University of Warwick, UK.

7   CEDIP Thermal Imaging Infra Red Array Camera (CEDIP SA, Croissy, Beauborg, France).

8   **Sharples, K.** (1981) *Oblique Incidence Device* (Sharples Photomechanics Ltd).

Case studies in rapid prototyping and manufacturing techniques – flow visualization using rapid prototype models

Chris Driver
Rover Group, UK

1 THE DESIGN PROCESS

Within Rover Group Power Train there is a continual effort to improve the quality of our products in terms of their quality, reliability, fuel economy, and emissions to the environment. In order to help design and development engineers to effect these improvements new analysis tools need to be developed. There is also a requirement generally within industry to shorten product development cycle times and costs bringing better quality products to market place sooner thus improving company profitability and competitiveness. Within BMW and Rover Group Power Train there is a drive towards reducing the number of working prototypes built within an engine development program (4). This is to be achieved by increasing the amount of both computational and experimental analysis performed early on within the program.

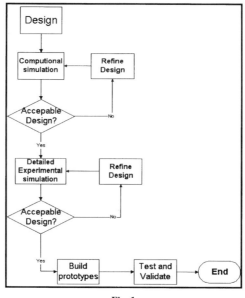

Fig. 1

Figure 1 shows how by using a greater number of design/validation loops within the design process product designs can be validated long before prototypes are manufactured. Over the past few years finite element (FE) modelling has been used to great effect to approximate stresses and fluid flows within components; however, with large and complex systems FE can be limited in the detail of information that it provides. With the use of physical experimental analysis in conjunction with FE far more detailed information about the performance of components and systems can be obtained, giving the confidence in designs to reduce the number of prototype build and test phases within a product's development. Although this has the effect of increasing the product development cost early on in the design cycle it means the overall products can be developed more cheaply and quickly (Fig. 2).

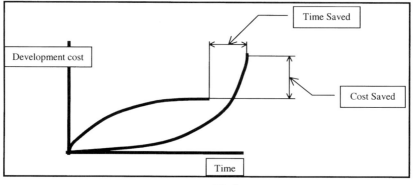

Fig. 2

2 FLOW VISUALIZATION

Flow visualization is a technique, in use within Rover over the past four years (2, 4), which can be used to determine very detailed information on the flow pattern of liquids or gases within an engine. The technique uses transparent models of parts of the vehicle, e.g. cylinder head and block cooling passages. A fluid matched for refractive index to the transparent model material and seeded with minute particles is then pumped through the model. This means that light passing across the surface between the test model material and the test fluid is neither reflected nor refracted allowing perfect optical access to the internal fluid flows.

Provided the outside surfaces of the model are at right angles to each other, these minute particles within the flows can be illuminated with a sheet of laser light giving an undistorted and very graphical real time illustration of the fluid flows within the model.

The laser light is delivered to a probe by fibre optics, allowing complete flexibility of viewing throughout the model (see Fig. 3). Detailed observation of the flow can show up areas of stagnation and turbulence.

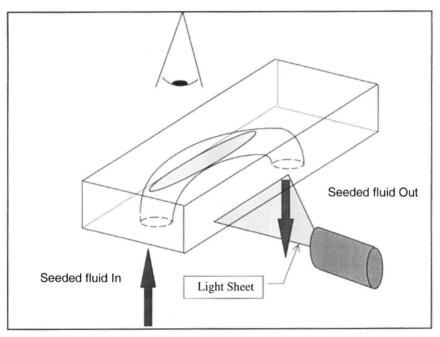

Fig. 3

3   FLOW MEASUREMENT

Particle image velocimetry (PIV) is a technique whereby velocities of fluids can be measured by photographing the same minute particles within the flows in two positions a short time distance 't' apart (1, 6). This is achieved by running the model as for flow visualization but instead of using a continuous sheet of laser light, the laser is pulsed twice a known distance in time apart as each image is acquired, as shown in Fig. 4. Storing these images as PC graphics files, specialist software is used to identify the particle pairs and plot velocity vectors for each particle movement.

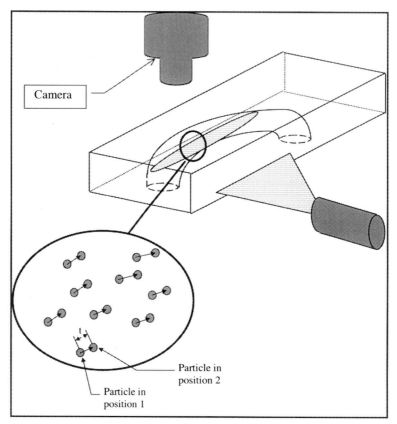

Fig. 4

4   CASE STUDY

The case study shows an analysis of part of an engine coolant circuit. A solid CAD model of the water passageway surfaces required is separated from the main components (Fig. 5).

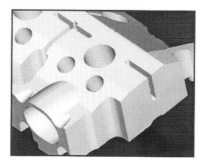

Fig. 5

 Next, still within the CAD environment, a thin layer of material is modelled around the outside, completely encasing the surfaces. The water passageway geometry can then be removed leaving a CAD model of a thin shell having the required geometry on the inside (Fig. 6). Any other features required can be incorporated, such as fixing holes, gasket faces, or supports for making mould boxes used later in the process.

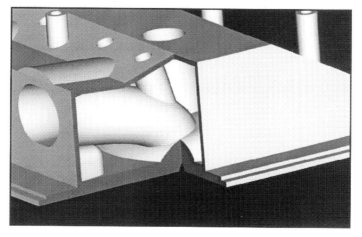

Fig. 6 (to view this figure in colour, go to colour section)

This CAD model can then be converted to an STL file and built as an SLA model. Attention must be paid to the orientation of the layers within the SLA model in relation to the flows that need to be viewed. Since the SLA material is not completely optically homogeneous, it is possible only to view across the layers while shining the light sheet in the plane of the SLA model layers. In the case of a cylinder head model, most of the flows of interest were horizontal so the layers of the SLA model were aligned in a horizontal plane and the model was viewed from above.

A box can then be built around the SLA model into which clear epoxy resin can be poured. The resin is then carefully set off leaving a clear block with flat sides having the water passageways inside. All that then needs to be done is for the outside of the model to be polished in order to improve the clarity of the surface leaving the finished model as shown in Fig. 7.

Fig. 7

A cylinder block coolant jacket model was produced using a similar method and the whole model mounted and run in a similar way to that described in Sections 2 and 3.

5   ANALYSIS RESULTS

5.1   Flow visualization

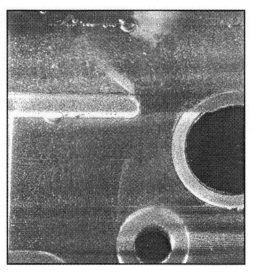

Fig. 8

Figure 8 shows an area of coolant flow from a section of the cylinder head around the spark plug. The coolant is flowing from the bottom of the image to the top and shows the effect of the rib pushing the coolant around the spark plug boss.

5.2   Particle image velocimetry (PIV)

5.2.1   *Raw PIV image*

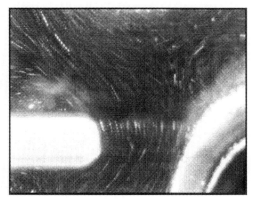

Fig. 9

Figure 9 shows a PIV image of the same area of flow but looking more specifically around the end of the rib.

5.2.2 *Processed images*

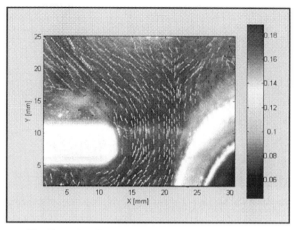

Fig. 10 (to view this figure in colour, go to colour section)

Using a software package called APWin (1, 5) it is possible to identify particle pairs within the image. The software works by identifying all of the particles within the raw image and calculating their centres of mass. It then takes each particle in turn and searches within a user-defined area for a second particle with which it can be paired. Having identified the second particle in each pair the software will output an ASCCI file having the 'X' and 'Y' positions of the first in each particle pair and the velocity and direction of the second. These ASCCI files may be displayed using a graphical analysis software package such as MATLAB as a vector map, as shown in Fig. 10.

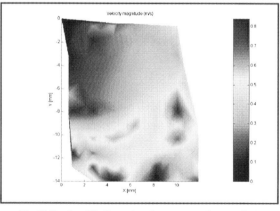

Fig. 11 (to view this figure in colour, go to colour section)

PIV data is dependent on there being an even spread of particles throughout the flow field at the time of the camera exposure. Where the data points are not evenly spaced across the image it may be useful smooth the results as shown in Fig. 11. This shows a Delaunay plot of the original vector map again produced using MATLAB. In this technique the software takes the original data points and plots extra points in between before displaying the data as a coloured contour map.

REFERENCES

1   **Udrea, D. D., Bryanston-Cross, P. J., Driver, C.,** and **Calvert, G.** (1997) Application of PIV (particle image velocimetry) and flow visualisation to the coolant flow through an automotive engine. European Symposium on Lasers and Optics in Manufacturing,

2   *Optics and Vision on Manufacturing*, Frankfurt, Germany.

3   **Calvert, G.** (1994) Flow visualisation. *Rapid News*, Spring Issue.

4   **Calvert, G., Driver, C. M. E,** and **MacDonald, J. A.** (1996) Rapid experimental analysis of flow. TCT conference.

5   **Calvert, G.** Rapid prototyping helping flow visualisation at Rover. Annual Conference of the Society of Experimental Mechanics, Dearborn, USA.

6   **Judge, T. R.** Qualitative digital image processing in fringe analysis and PIV.

Section 2

Utilizing Bureau Facilities

Overview of utilizing bureau facilities

Chris Ryall

There are currently two options for those wishing to utilize rapid prototype models. Purchase a model from a bureau or purchase a machine. This sounds quite easy but the reality is not quite so simple. There are many logistical problems, which while not very difficult to resolve, can easily cost time, money, or at least cause embarrassment.

The provision of good data is usually the first stumbling block. In the early days of rapid prototyping a large proportion of all files had faults, which caused considerable problems. The reality is that poor CAD or STL file translation results in poor models or disappointment. This, though, is not the subject of this section and is covered elsewhere. But it is a problem that needs to be explored before embarking on either route. The easiest solution for those wishing to use a bureau is to provide a file to a bureau as a trial to gain necessary confidence. The chances are that there will not be a problem, but if there is then the bureau should be able to provide an indication of how to overcome it. Unfortunately, the situation becomes more difficult where CAD data are being supplied from a range of CAD systems. At a meeting recently, two companies stated that of the many thousands of files they received each year 75 per cent were perfect, a further 15 per cent required some sort of repair before they could be manufactured, and the final 10 per cent had to be referred back for some remodelling.

CAD problems can cause major difficulties for RP users, particularly if they are relatively inexperienced. Delays and machine down time due to bad CAD data is a factor that may not have been considered in the business plan. It should be hoped that, in the future, more robust CAD packages and better training would resolve all CAD issues.

The following papers cover some of the eventualities and consideration of opting for purchase of equipment or using bureau services. In the first paper, Peter Bessey covers the issues surrounding the use of a bureau. It is important to realize that all bureaux are not the same, which can have major implications in terms of cost, quality, and the service they provide. Ian Halliday's paper discusses the relative possibilities and implications of running an internal prototyping bureau. The paper has been written from the perspective of a large automotive company, but many of the points raised are applicable to both large and small concerns.

Using bureau services

Peter Bessey FCSD
Senior Partner, Hothouse Product Development Partners, London, UK and Hon. Research Fellow, London Institute/Central St. Martins College of Art and Design, UK

SYNOPSIS

The advent of new technology continues to reduce prices, but, with few exceptions, the cost of equipment for rapid prototyping remains substantial. Many designers and engineers can benefit from these technologies only by using outside agencies. Few have the luxury of in-house facilities of this sophistication, unless part of a substantial company, and even fewer have access to a wide range of methods. Bureaux are therefore the choice for many users. But buying these services successfully first requires an awareness of their differing advantages and, what is more important, of their limitations. This section does not seek to describe the various methods in detail, but to highlight some fundamental issues arising in their use and to provide a basis of understanding from which a selection can be made.

1   IDENTIFYING A SOLUTION

1.1   Firstly, what are your aims?
The golden rule with most things computer-related, is to first define your objective, and then to identify what will best meet your need. Only then should quality, delivery, and price be considered in selecting the right supplier. That principle holds good for rapid prototyping bureau services too.

Rapid prototyping exists in many guises. More will arise as these useful and powerful technologies develop further, together with the materials they employ. So it is important to select the right solution for your purpose and that will vary from project to project and from component to component.

1.2   What factors affect your choice?
These will certainly include:

- model material performance;
- model durability;
- component size;
- dimensional accuracy and resolution of detail;
- degree of difficulty of form;
- quality of finish;
- quantity of repeat parts;
- availability of additional processes;
- cost;
- geographical location;
- data transfer compatibility;
- time scale;
- supplier reliability.

Compare your purpose in making the model or prototype against these factors, and establish priorities before seeking a supplier.

1.3 Some limitations

Many bureau sources for rapid prototyping will offer a single solution, which may be suited to a specific industry. Others will have a mix of methods. Few, if any, will offer every possible type of process and material. You may not be able to obtain impartial advice on what best suits your case. Bureaux, after all, have a commercial interest in exploiting their own resources.

Most suppliers can deliver models at different levels of finish to external or internal surfaces. Before placing an order, you should ensure that you fully understand what these levels actually represent in relation to your model. This will both avoid unnecessary cost and assist a correct result.

Having identified which process and finish quality best match your requirement, do not be persuaded to use the most readily available, just because of convenience, unless that really is appropriate. What is delivered could fall far short of your actual objective.

Why is this? Experience shows that disappointment arises primarily from the real limitations of each technique in areas of dimensional accuracy, detail resolution, and material performance. It may also arise from the quality of service and output provided by the bureau or from a difference between uninformed expectation and what can realistically be achieved.

2   MATCHING PROCESS TO PURPOSE

Each method creates prototypes that possess different characteristics. Your original purpose for the model should drive your choice.

That purpose may be for a straightforward validation of the CAD model data; or it may extend to cosmetic assessment of the overall design; or for some level of functional trial.

2.1 Which purpose?

If for CAD data validation, then the model needs to be sufficiently stable for the appropriate degree of dimensional checking, handling, and for fit against other components.

If for cosmetic assessment, then all aspects affecting an aesthetic judgement must be catered for. To what degree is resolution of surface detail, finish, and accuracy of component form important? Will the ability for the model to be handled or to accept alternative coloured coatings be required?

If for functional evaluation, then all necessary practical aspects of the component must be considered. To what degree will material performance, finish, and dimensional accuracy contribute to a validation of the design, the component fit, or to movement of the part? Will the model survive extremes of temperature or other necessary environmental conditions of use?

Where large parts are in question, will it be necessary to manufacture these in sections, with consequential visible joints and functional weaknesses?

Are multiple replicas of the part required for wider evaluation or demonstration? Are they to be handled with care by technical staff or roughly by unknowing personnel? Will the first component be needed as a pattern for casting or mould manufacture? If so is shrinkage to be incorporated into the model and how can possible distortion be avoided? Will the part survive for additional uses?

What is important for orientation of the part during the build process? Where should supports be attached?

3 CHOOSING THE PROCESS

This should be dependent on the match between the advantages and limitations of each RP process and the requirement of your particular component. There are various factors that may influence your choice. A summary of major issues is provided, which covers the most commonly found technologies. Similar considerations should also hold true for governing the selection of new techniques as they emerge in the future.

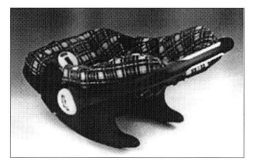

Fig. 1 Infant seat full-size SLA

Fig. 2 Infant seat full-size SLA

3.1   Stereo-lithography (SLA)

This process is probably the best known. There has been remarkable improvement made to the durability and choice of resin materials and the thickness of layers used to build models. Materials now exist with specific properties such as rigidity, flexibility, high-temperature resistance, and optical clarity. But models require a degree of hand finishing in order to remove residual surface steps, which can be difficult within small cavities. Functional use of the models is limited by the nature of the materials available. Parts can be used as pattern masters for replication in secondary processes.

3.2   Stereo laser-sintering (SLS) and direct metal laser-sintering (DMLS)

Now becoming more common, this process permits model building in a limited range of plastic polymer and metals. Surface quality and detail resolution are somewhat reliant on the particle size of the powders used in the build. That is improving, but additional finishing is necessary to provide other than a textured surface. Model parts can be used functionally and as masters for replica casting. The metal powders offer a means of creating tooling for prototype injection moulding and die-casting runs. In general the metal component builds require further processes which can include oven baking and impregnation, with consequential shrinkage and time delays.

3.3   Fused deposition modelling (FDM)

Not widely available as a bureau service, this process builds using wax, rigid plastic polymer (ABS), and elastomeric material. The models can be used for quick visualization of parts, as replication masters, or directly as patterns for lost-wax casting. Hand finishing is required to remove surface steps. Where the build polymer suits, functional use of the models can be achieved.

3.4   Three-dimensional printing and solid object printing

Again not widely available from bureaux, the two techniques utilize thermoplastic waxes or polyester resins, suited to the building of model visualizations, replication masters, or lost-wax patterns for casting. The wax material can provide very fine detail for small parts and has been used in jewellery development. The process also permits rapid manufacture of wax mould cavities for resin casting in a more durable material that can simulate production polymer.

3.5 Laminated object modelling (LOM)

One of the earliest techniques, but not widely seen at bureaux, this method is often applied to large models that require particularly robust properties. Surface finishing by hand is needed to remove layer steps. The models can be used for visualization, replication masters, and patterns, particularly for sand casting. A variation on this method permits manufacture of low-cost, hand-assembled plastic parts using sheet-cutting technology, but necessitates considerable hand finishing to conceal or remove stepping.

3.6 Computer aided machining (CAM/CNC)

Often neglected as an RP process, this can be the most valuable method of producing single parts or a small batch in production-intent material. Its form and detail limitations are governed by cutting tool diameters and the number of axis required by the model. Machined block material will not provide an identical performance to that of cast or moulded components, due to flow, grain, and skin effects, but does allow close approximation in many cases.

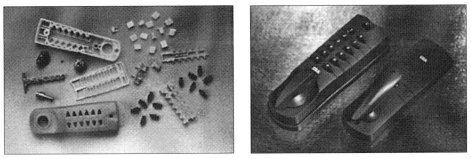

Fig. 3 Digital lock components CNC   **Fig. 4 Digital lock models CNC**

Supplier choice may be limited by their machine capability. This will control the number of machining axis, throat, maximum table size or travel available.

As an alternative to milling, high-speed routing should offer a cost-effective solution for softer materials, which can include aluminium. Surface finish quality will be controlled by tool and cutting speeds together with cutting path frequency. Not all components or detail features will be suited to this method due to restricted tool access for particular forms.

3.7 Rapid tooling

Becoming increasingly available and using a widening variety of tooling materials and processes, this is the only method that can easily provide a volume of moulded or cast components in production-intent materials. Dimensional accuracy, repeatability, and the maximum number of produced parts will be variable, dependent on the 'hardness' and stability of both the tooling and the moulding material. This factor and the associated cost and time scale will influence choices. The available variations within this method require careful investigation before selection of an optimum source, but a full choice of production material should be possible.

3.8   Traditional model making and prototyping

Today, this method is not seen as 'rapid'. However, when combined with other RP systems, the mix may provide the only economic means of achieving your needs, and modern practices linked to CNC can sometimes outperform standard RP techniques. While this approach will not automatically validate CAD data, if applied intelligently it will act as a good 3D check. Prior to CAD/CAM it represented the only method available and can still compete today – but success is reliant on proper control of the human factor.

4   EXPECTATIONS AND DELIVERABLES

For finely detailed cosmetic parts, the potential for variations added by a need for hand finishing, can lead to unacceptable results. Similarly, where accurate function and fit require additional machining processes to provide essential precision detail, difficulties may emerge. If these extra tasks are carried out with any lack of sensitivity or with insufficient knowledge of what the customer is seeking, an operator can destroy or diminish the very features most important to a component design.

This is where a bureau's geographical location becomes important – a personal visit, explanation, and supervision of the work can assist in delivery of the right result. It is perfectly possible to obtain satisfactory services over the Internet, but when deadlines and critical projects are at risk, it is best to rely on previously proven and easily supervised sources.

The validation of a close-tolerance assembly is not ideally suited to any of the processes that require hand finishing. In these cases, CNC machining or rapid tooling methods are more appropriate. Material properties that can be delivered only by the actual production process will dictate the use of a prototyping technique, capable of replicating them.

As in most fields, there is a wide variation in both the content and quality of services available from independent commercial bureaux, which requires the same cautionary response in selection. Resource can also be obtained as 'over-spill' from other less commercial establishments. These include autonomous institutions such as universities, colleges, and research organizations, together with certain access to the in-house departmental facilities of some major manufacturing companies.

When choosing to source urgent work from non-independent organizations, it is wise to confirm what level of priority you have on their schedule and to check that 'state-of-the-art' technology is being offered. Remember the sense of urgency and the rapidity common to your industry may not be understood elsewhere. Make certain that it is.

In all cases it is worth meeting with your supplier to discuss your needs in the first instance, to view their facilities and to assess their competence in relation to your project. Do not hesitate to seek reference customers and examine samples of recent previous work.

Ensure that you understand the bureau's own specification for levels of finish and be familiar with the availability of other on-going processes, such as machining, traditional model making, painting, vacuum casting, silicon rubber, or rapid tooling services.

Prove out data transfer methods prior to real-time use. That will allow resolution of any problems before deadlines start to bite.

Consider your time scale and the practical limitations on a bureau to deliver to your specification. Measure cost quotations against service level and reputations.

When you eventually commission and receive the model, satisfy yourself that all is in order and, if there is time available in your project schedule, ensure that any defect is corrected before invoicing and payment take place.

If you have taken care to ensure that your original instruction and specification is clear at the outset, any potential discrepancy or conflict with your supplier will be easier to resolve.

Finally, once you have established a good working relationship with a reliable supplier, do not switch to others without good reason. You may not automatically receive the right result without proving the process all over again.

Running an internal rapid prototyping bureau

Ian Halliday
Chief Engineer of Rover Group RP&T, UK (1994–2000)

1   INTRODUCTION

With so many highly capable rapid prototyping bureaux available at an ever-decreasing cost, what is the purpose of having an internal rapid prototyping and tooling capability? This paper aims to answer that question by looking at the purpose that the Rover Group facility had and its interaction with its external counterparts. Although the Rover Group facility has now taken on a new form since the sale by BMW Group, many of the principles and thoughts outlined remain true in the view of the author.

The article puts forward the case that the rapid prototyping and tooling (RP&T) function should lead the way with these technology sets. However, there are other highly successful RP&T capabilities in other automotive manufacturers that take a 'fast follower' approach. There are advantages and disadvantages to each approach. The technology leader can take advantage of new technologies earlier and potentially drive integration forward more quickly. The fast follower is likely to make fewer mistakes by observing the leaders but pays the price of later take-up and integration into the business. Each company should match the approach chosen to the needs and nature of the company.

Many large companies now have an internal bureau for the rapid production of prototype parts. In Europe, almost all of the major automotive companies have a significant prototyping capability. The automotive sector is not the only manufacturing sector to do this, of course, but the continuous drive for shorter design cycle times has led it to use rapid prototyping processes aggressively. Some automotive companies invest in new technologies at an early stage in their development in an attempt to gain a competitive advantage; rapid prototyping is just one example.

2   THE ROVER GROUP RAPID PROTOTYPING AND TOOLING (RP&T) FACILITY

Rover Group had an internal rapid prototyping capability from 1991, when the purpose of the original facility was primarily to evaluate the new stereolithography process. The overall objective was to find practical applications for the new technology and those that followed.

The need to maintain a practical focus naturally led to an interest in rapid tooling and rapid casting. Consequently, in 1994, it was decided to develop the team to include these technologies and forge the partnership with Warwick Manufacturing Group (WMG). Over the next three years the team grew from seven to 34 people and became strongly integrated into the business.

In 1998, the Rover Group RP&T facility produced 2200 SLA parts, 3200 SLS parts, and 9500 polyurethane parts from silicone moulds. In addition, 200 castings and 3800 parts from rigid rapid tooling were made for a range of internal customers to excellent effect. In the same time span, the Rover Group team worked on 14 research projects in conjunction with WMG. Overall, the Rover Group RP&T facility saved the company approximately £3.5 million in 1998. Also in 1998, the Rover and BMW Rapid Prototyping facilities were joined together with the aim of improving the BMW focus on rapid prototyping in Germany.

'Rapid prototyping and tooling' is gradually becoming 'rapid manufacturing', having changed in a very dynamic way over the last ten years. Equally, the relationship between the internal and external bureaux has changed. Therefore, before reviewing the purpose of an internal rapid prototyping bureau, it is worth looking at the alternatives for prototype part supply from the large company viewpoint.

3   THE ALTERNATIVES TO AN INTERNAL BUREAU

3.1   External bureaux

When Rover first investigated stereolithography in 1989/90, the very first (fragile!) part was purchased at a relatively high cost from a bureau (that no longer exists). The fact that it was possible to try out the new technology by this route made the bureau indispensable. No investment was necessary, just a small amount of risk taking.

New technologies are usually expensive and risky to invest in, so the output is costly to the consumer. Consequently, that first part was very expensive because the new technology could command a premium price. Nowadays, a rapid prototype model is more or less the norm and so the profit margin is lower.

Sadly, businesses in a rapidly growing high-technology marketplace can come and go. Building a long-term relationship with an external bureau may not be that easy when dealing with high-investment, high-risk technologies. Large companies like to build up 'preferred suppliers', not easy if the suppliers disappear like that first rapid prototyping bureau did.

Now that rapid prototyping (RP) is more stable, the story is very different. External bureaux can provide a very fast, reasonably cost-effective, efficient service, with no risk to the customer. So when it comes to getting parts through the system, the external bureau is very attractive when the technology required is not quite so new.

3.2 Reliance on production suppliers

Many companies can be slow to pick up on new technologies if it is outside their experience. Although bureaux can help to accelerate the process by providing easy access to the new technology, the large investment involved in setting up an internal bureau may take a long time to get through the company system. The result is that new and costly technologies are often only adopted once they have developed. Hence, even the largest and most competitive automotive suppliers may have only had an RP capability for the last five to six years. Suppliers to the automotive industry have often been slower to take on board rapid prototyping. A pure reliance on the supplier base may therefore result in years of lost learning about how to use new technologies such as RP to best effect.

Automotive companies cannot afford to wait until the supplier base has caught on to new technologies. An automobile manufacturer must get in early, experiment, learn quickly, and then maximize the exploitation of the new opportunities.

3.3 Have an internal team that simply outsource work

Porsche managed for a long time without having their own rapid prototyping capability. Despite this they were still among the best users of stereolithography and laser sintering during the early 1990s. The reason for this apparent anomaly was that the Porsche Motor Sport design team made a determined effort to keep up to date with the developments in rapid prototyping and exploit them.

Although their 100 per cent outsource approach was very successful, the Porsche team were determined to get their own internal equipment due to the extra speed, cost effectiveness, and learning that they would gain from it. In 1998 they succeeded and have continued to build their capability since. They proved that it is possible to be very successful with a capable and determined RP outsourcing team but that a dedicated internal RP facility was better for them.

4   THE FUNCTIONS OF AN INTERNAL BUREAU

4.1   Make prototype parts

At first, it may seem that the only function of an internal bureau would be to make parts for the company concerned. Simply making parts is only one aspect of the story though; the biggest potential benefits come from having an unbiased and competent internal prototype parts manufacturing team.

The internal bureau must be competitive in terms of speed and quality with the external bureaux. The speed of service should really be better due to the infrastructure of the company, logistics becoming an increasingly important issue as RP build times decrease. However, it can easily happen that the internal bureau becomes overloaded with work, resulting in a slower response time. The external bureaux often have the advantage of having many machines available for use and so can respond very quickly to a request, e.g. Materialise 'Next Day' service. For that reason the internal bureau must work in partnership with external bureaux for maximum efficiency.

By outsourcing overload demand to bureau, the internal bureau can focus on higher value-added parts that require higher secrecy, quality, or the use of one-off special processes. The aim is that where a premium price might be charged by an external bureau, the internal bureau should be the first choice. In most cases, a new technology demands a premium price.

The internal bureau should work in partnership with external bureaux, building up strong working relationships with a small number of them. It can be a symbiotic relationship, benefiting both parties. By managing the outsource of prototype parts, the internal bureau can use the knowledge they have to best effect. The external bureaux can also gain by reducing their administrative effort through having 'preferred customers'.

Probably the best reason for making prototype parts internally is to learn from the practical experience and then use that knowledge to benefit the company. In effect, the small amount of prototyping work done internally can be used to 'leverage' much greater benefits by putting the knowledge to best use. Making parts is therefore only the beginning of the story. The real issues of interest relate to the identification and exploitation of new prototyping technologies.

4.2   Identify new processes and technologies
To make best use of the new technologies available, you first have to know what they are and why they might be beneficial. In the Rover and BMW internal bureaux, the task of learning about new prototyping technologies was taken very seriously. It is a continuous proactive process and includes:

- involvement in conferences (both passive and active);
- working on European and UK projects;
- involvement in the technical community;
- visits to other companies around the world.

The overall aim of this activity is to ensure that new opportunities are not missed due to ignorance of their existence. Early awareness helps to ensure that new technologies can be tracked during their early development and then adopted by Rover and BMW at the right time.

4.3   Make new technologies and processes viable
Every bureau, whether internal or external, has to work with each new technology as it is adopted to ensure that it is implemented to best effect. The fastest way to do this is by having a specialist group such as the one at Rover's facility in the UK. The Rover new technology implementation team is further strengthened by the partnership with WMG. The partnership enables a wide range of technologies to be worked on, from those in the early stages of development to well-established ones.

The specialist new technology team can focus their efforts on making technologies meet the specific needs of the business. New rapid manufacturing technologies are usually generic and great improvements can be made by tailoring their functionality to meet company demands. For example, Rover Group was able to fine tune the use of laser sintering materials for the investment casting of highly complex cylinder heads for development engines.

Other important development options include running Beta trials on new materials and processes. Rover found the structured nature of the Beta programmes very useful in gaining rapid early learning about a new material or process.

4.4 Integrate new processes and technologies into the business

It is obvious that a new technology is of no use unless it is used! The reality is that people will not use a new opportunity if they do not know about it or cannot see how it would benefit them. Taking the argument to its full extent, until a new technology is fully integrated into the business processes, it is not fulfilling its potential. The quicker the new technology is integrated into the business, the longer its useful life will be.

To drive forward the integration of a new technology, the internal bureau must perform a marketing role. The task includes:

- creating awareness;

- educating engineers about prototyping opportunities;

- providing samples for trial;

- highlighting potential new opportunities;

- being clear about risks relating to the new technology;

- background information about related technologies and processes;

- generating case studies.

There are very many ways of 'advertising' a new process, all of which requires a significant amount of effort. Although many suppliers and bureaux advertise their capabilities within large companies, they tend to focus their efforts on certain groups or on certain technologies. An advantage of the internal bureau is that a broader approach can be taken, covering the whole engineering community using a wide range of technologies and processes in an unbiased way. This neutrality is the basis of the role of the internal bureau as a company 'champion'.

4.5 Be company champion of best practice

The first role of company champion for rapid prototyping is to identify and make up for the gaps in the technologies being used by the business. Because not all the suppliers have access to state-of-the-art technologies and processes, it is advantageous to have a group whose job it is to identify the gaps and help to bring up the average standard. In this way, the relatively small direct output of the internal bureau can have a large beneficial 'leverage' on the whole business.

The company champion role can be part of the marketing function of the internal bureau. Information can be supplied to company or supplier engineers about the latest technology options and their capabilities.

Another role is to directly build up the capabilities of the supplier base. For example, the rapid casting capability at Rover has involved a lot of work by the Rover Group team directly with interested and capable foundries. The result has been beneficial to both Rover Group and to many other businesses in the UK.

The internal bureau can co-ordinate actions within the business in a way that could not be done by any external group. Standards and common understandings can be gained that are beneficial across the whole business. It can also build up a technology portfolio that is designed to match the company's needs.

The company champion role represents the overall purpose of the internal bureau in many ways – such as creating awareness of new opportunities, by ensuring that the company can take advantage of every suitable new opportunity, by providing an unbiased and balanced view of future potential and by understanding both the new technologies and the internal needs of the business.

5   SUMMARY

The internal bureau is an essential part in the picture when it comes to the use and supply of prototype parts within a large company, working in partnership with external bureaux and suppliers to help to optimize the supply of prototype parts.

Making prototype parts is only one aspect of the role played by the internal bureau. To make new technologies and processes perform to the best benefit of the company, a lot of development work has to take place to tune them for business use.

Once new technologies are ready for use, a key task is to integrate them as quickly as possible into the normal business processes. In a company the size of Rover or BMW, the integration task is huge and takes a lot of effort over many years. However, the quicker the integration can be achieved, the longer the effective life of the new technology investment.

The internal bureau is in effect a company 'champion' for best prototyping practice. Performing this role to full effect means working in partnership with suppliers and external prototyping bureaux. The goal is to establish a framework that enables the company to increase its competitiveness through the use of new rapid manufacturing technologies. The co-ordinating 'champion' role of the internal bureau accelerates the rate of improvement of not only the host company but also the bureaux and supplier partners. In this way everyone gains from the existence of an internal bureau within a large company.

Section 3

Rapid Casting Techniques

Overview of rapid casting techniques

Chris Ryall

Since rapid prototyping (RP) first gained general acceptance in industry, users of the technology have looked for applications to enhance the usefulness of what was originally an expensive three-dimensional (3D) copy of a CAD model. One of the first application areas was casting. These early projects were sometimes quite risky, as the effects of binder systems and processes were unknown on the models. A few foundries were very successful, consistently producing some excellent castings. Others were not so fortunate, resulting in a high degree of dissatisfaction. Since these early days the manufacturers of RP equipment have developed newer materials, achieving greater levels of accuracy, tailoring them to the various casting processes. Foundries now have a greater awareness of the different techniques and how to handle and process the different types of model, and in some instances operate their own RP equipment.

Probably the most widely used application for the generation of castings from RP models is in investment casting. The main reason for this is that no tooling is required, regardless of complexity, for the manufacture of the casting. Typical turn around is widely reported as being anywhere between 1 and 2 weeks, and in some instances less.

Investment casting involves encasing a sacrificial RP master pattern in a ceramic mould. Once the ceramic has solidified, the master pattern is removed by autoclaving or firing of the ceramic mould. A metal is then poured into the resultant cavity, generating a metal clone of the original master pattern. The direct investment casting of RP patterns is widely used but the process does not guarantee a sound casting, as assurances cannot be given that there will be no problems when creating the mould or casting the metal. There is therefore a risk, albeit very small, associated with the production of castings via this route, or for that matter any other casting route.

Fig. 1 Sacrificial RP pattern

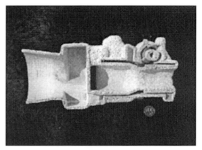

Fig. 3 Ceramic shell with RP pattern removed (sectioned)

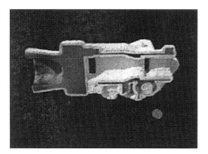

Fig. 2 Shelled RP pattern (sectioned)

Fig. 4 Investment cast metal component

(Photographs courtesy of Land Rover Limited.)

Unfortunately, the direct use of RP models in investment casting does not provide a universal solution for the production of prototype castings. Where higher numbers of castings are required the route may become too expensive due to the cost of each pattern. There are other processes, which do not involve the same level of risk, primarily because tooling is used allowing the generation of multiple moulds or patterns.

Components with complex internal cavities are not necessarily suited to investment casting, due to difficulties in the removal of the ceramic material from internal cavities. Typical examples of such cast components are cylinder heads and hydraulic valves, which have traditionally been sand-cast. Another issue that may necessitate the use of an alternative process is the need to encourage the correct grain structure within the casting to obtain the desired mechanical properties. In these instances, only the intended final production process for the casting may be used. One such process is pressure die casting, where a steel die is required. Currently, the manufacturers of sintering machines are actively exploring the generation of die casting tools by direct and indirect means. Alternatively, a tool can be cast from an RP pattern, which in turn can be used for the manufacture of die cast components.

With all RP techniques there is usually a limitation to the process, but as time moves on these limitations may well be reduced or even resolved. It is therefore important when considering the use of any RP technique to understand the advantages and disadvantages of each process before selecting one for a project.

The following Section addresses the issues raised above. Craig Vickers of IMI offers a solution for low-volume/prototype investment castings. Pete Harrison of WCM Patterns looks at the application of RP patterns to a range of casting processes and Anders Karlsson discusses the effects of rapid prototype castings using laser sintering of sand on the design cycle. In the final paper Greg Redden describes the results of a project looking at problems associated with producing prototype diecast components with representative mechanical properties. It should be remembered that these techniques are additional and complimentary to traditional practices and should be used appropriately.

An alternative route to metal components for prototype and low-volume production

Craig Vickers
IMI Rapid Prototyping, Birmingham, UK

1   INTRODUCTION

Much has been learned and even more written about the ability and method of producing metal components using various rapid prototyping (RP) techniques. Most people with some knowledge of the RP industry will probably have heard of Quick-cast SLA, Tru-form SLS, Cast-form SLS and other processes using the variety of Thermojet printing machines now available. Each of these routes and any others of which the author may be unaware, have their advantages and inevitably their disadvantages for the production of metal parts via investment casting. The one common factor in all of the above is the fact that they use an RP process to produce a pattern or patterns which are then sacrificed during the casting process. This can prove costly if a number of castings are required.

In the UK, there are hundreds if not thousands of investment casting foundries. Many of these foundries have developed the ability to use RP patterns within their process, indeed some have even invested in RP machines to supplement and enhance the services they offer. However, the one thing most, if not all, would admit is that they have far more experience and expertise in producing castings from wax patterns than from RP masters. These foundries all use waxes every day and would prefer wax to RP master whenever possible. During our considerations for the route to metal we would offer to our clients, this was one of the major factors taken into account.

IMI Rapid Prototyping was formed in 1993 and while this is not intended to be a 'sales pitch', it is important to give some background as to how we have developed our methods and also, some of the thinking behind it. In the relatively early days of RP in the UK, we were building SLA models and creating silicon rubber tools from which we could vacuum cast plastic and rubber parts. We were, to a certain degree, learning as we went along. Inevitably, after a short period of time, the first requests came for metal parts. These were initially declined but it soon became apparent that there was a demand in this area and, consequently, some low-level development work started. It did not take too long to realize that it was possible to pour wax into silicon tools created from SLA models: there were lessons to learn with regard to contraction of wax, and allowances to make for this and the contraction for metal, but these were really just exercises in maths. We were soon in a position to talk to foundries about producing castings.

Discussions were not always easy since we were always talking about low volumes and inevitably short delivery times, but we were at least aided by the fact that we were taking along waxes and not something the foundries had not used before. This allowed them to process our patterns alongside their own with a minimum of disruption to their business. Initial results were promising and development continued. Eventually after much trial and error, we were able to define the wax production process from our silicon tools to a degree where we were able to extend tool life from an initial 20–30 waxes per tool to in excess of 100 waxes. This has since developed further and, as will be shown later in one of the case studies, the record now stands at 1400 waxes from one silicon tool.

There remained one major problem in the production of metal parts as far as we were concerned and that was the ever-present issue of delivery time-scales. We were always fighting the fact that although we were sending waxes to foundries, they were in low volumes and represented a tiny percentage of the business of the foundry. Typical time-scales as recently as 4–5 years ago read: SLA models 3–4 working days; wax patterns 4–5 working days; investment castings 4–5 working weeks. For customers who were becoming used to faster deliveries offered by the RP processes, this was not what they expected.

How could we as a company tackle this issue? The quantity of waxes we were able to produce from our tools was increasing and offering some promise, but without cracking the delivery problem we could be wasting our time. A great deal of time was spent talking to various foundries and trying to find a solution, it became very obvious that the RP industry requirements, and by this time, we as a company were regarded as a disruption and a minor irritation rather than an opportunity. Fortunately, we were able to convince some foundries of the potential we saw and managed to get their agreement to attack this area together. We outlined three things that we needed from these foundries. These were (i) fast responses for quotations; (ii) fast turnaround of castings; and (iii) competitive price levels. Quality was taken as an obvious requirement and could not be compromised.

One foundry in particular took up the challenge and we have now been working together for over five years. In this time we have progressed from being an irritation to being one of their major customers. Due in no small part to their commitment and ability, we have been able to push up the quantities of metal parts we supply and as previously mentioned the quantity of wax patterns possible from our tools has increased dramatically with this. The important thing that we keep coming back to is the fact that we supply wax patterns. We can, and sometimes do, supply Quick-cast SLA models and the foundry will cast them, but they and we are happier to supply and produce from wax patterns.

There is always the danger when we talk greater quantities of castings that customers will think that if they only want one or two parts, they should still use the RP master routes. This is sometimes true but although RP prices have reduced significantly over the last five years, wax is always going to be cheaper than SLA resin or SLS powder. Of course, we must build a SLA or SLS model to allow us to create our silicon tools and these costs must be taken into account. We would always advise clients to compare costs of the different routes but also to take into account the risks associated with each route. Few foundries will guarantee 100 per cent success rate for conversion of SLA models to castings, however, we regularly achieve a scrap rate of less than 5 per cent of waxes supplied and also always supply excess waxes at our expense to allow for this. As mentioned previously, this is something we can do since wax is relatively cheap.

We cannot and would not claim that our method is suitable for all clients or applications. There are tolerance issues always to take into account and we freely admit that, in this area, some of the other processes are a better option since silicon rubber tooling and wax do add to the tolerance build-up. We recommend that for critical areas, a post-machining operation should be considered, but more importantly we advocate full and detailed discussion before embarking on projects.

Case studies are often a difficult area within rapid prototyping, since confidentiality is always of paramount importance. Below are brief details of three of the many projects supplied over the last few years. While these all involve larger volumes, at least two of them show examples of the benefits of having silicon tools, since quantities grew over and above the initial requirements. Also, it should be noted that while the examples below are all of aluminium castings, projects have been and are regularly undertaken in stainless steel, zinc, silicon brass, etc.

2   CASE STUDY – DYNACAST/FORD ALTERNATOR BRACKETS

We were approached by Dynacast Limited who had an urgent requirement from Ford to produce prototypes of two different designs of alternator bracket. They had been successful in gaining the production order for die-casting and were now tasked with producing metal prototypes prior to tooling. The sizes of the two components were both in the region of 300 mm x 300 mm x 200 mm, and material requirement was for aluminium. Parts were going to be LM24 material in production and therefore LM25 was required for investment castings.

3D CAD models in Pro-Engineer were provided and the requirement was for 50 castings of each within four working weeks. It was immediately felt that this number of parts was unachievable in this time-scale and we advised that we would supply 30 of each part in four weeks with the remaining 20 to be supplied one week later. We were authorized to proceed.

Three days after order placement, the SLA models were supplied for approval. Contraction for wax and aluminium had been applied but this did not prevent Ford from trial fitting them to units. The models were returned the following day with two minor modifications requested, which were carried out during the hand finishing operation. Parts were then set up for tooling and silicon rubber tools produced two days after return of SLA models. Tools were cut and proved the following day, these tools were relatively simple three-part tools and proving/venting of tools took place the same day.

Due to the size and volume of the part, it was only possible to produce six wax patterns per day from each tool, however, this had been taken into account during planning and our foundry had agreed to process the waxes in lots of 30. Consequently, five days after proving the tool, the first 20 waxes were shipped to the foundry, with the following 20 plus spares shipped five days later.

There were few problems encountered during casting. At quotation stage, the foundry had requested that we incorporate feed gates into the wax patterns and these had been added to the SLA model prior to tooling. By doing this, we were producing the waxes with the in-gates in place and therefore reducing the amount of work needed prior to casting. Within ten working days, fully heat-treated castings were returned to us for fettling and supply. Key areas had been identified and a post-machining and drilling/tapping operation was carried out on each casting before parts were supplied on time. Second delivery was made the following week and, as far as we were concerned, the project was complete.

Some three weeks elapsed before we were contacted by our client and informed that there had been a design change to one of the components. We were requested to produce a further 40 off castings to the new design. A new SLA model, tool, and waxes were created and the new castings were supplied after another four weeks.

At this point, production tooling was approved and started, however we were asked to produce a further 200 castings of each design in the interim. These were produced from the same silicon tooling and were supplied at the rate of 25 castings per week commencing 4 weeks from order.

The design change identified between first and second iteration of prototype had apparently been made after the prototypes had highlighted an area, which on Finite Element Analysis had been "borderline" for strength. Fitting actual parts allowed the customer to run these to failure and so prove or disprove the design prior to production tooling.

The prototype bracket is pictured below.

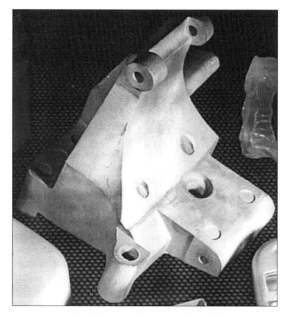

Fig. 1 Dynacast prototype bracket

3   CASE STUDY – SWITCH FINISHERS

A major automotive manufacturer approached us with a requirement for metal components to be fitted into 'crash vehicles'. These parts were the left-hand and right-hand version of a switch finisher and the dimensions were 130 mm x 60 mm x 60 mm. Parts were to be covered and therefore surface quality of the castings was not vitally important. The number of parts required at that time was not clear, but between 20 and 30 sets of parts were expected to be required.

From CAD data supplied, we provided costs and delivery times to produce these parts. An order was placed and the SLA models were taken for approval three days later. At this point, it was noticed that the parts did not line up with electrical contacts. The design was changed and a new set of SLA models produced and again taken for approval after a further two days.

On this occasion, approval was given to proceed with tooling, waxes, and castings. Component geometry was such that a relatively simple two-part tool was created for each hand. Wax production started three working days after SLA approval and with a component of this size, we were able to cycle the tool every 20 minutes, thus allowing us to produce in excess of 20 sets of wax patterns per day. Consequently, three days after production started, 40 waxes of each part were transported to our foundry for casting.

At enquiry, the foundry had confirmed that they were able to produce ten castings on each 'tree' and would attach in-gates themselves. Finished castings were supplied back to us 13 working days after wax supply. There had been a very low scrap rate and 38 castings of the right-hand part and 39 castings of the left-hand part were returned. There was no machining or heat-treatment requirement on these castings and, consequently, 30 pairs of castings were supplied to our customer the following day.

These castings were covered, assembled, and fitted to vehicles. In production, the intention had been to use pressure die-cast components and production-tooling costs had been obtained. The quality of the castings and the fact that they were covered led the customer to reconsider their position with regard to tooling. The total production quantity was estimated to be no greater than 1000 units and rate of vehicle build was well within the rate at which castings could be supplied via our methods. During this period of consideration, we were asked to supply a further 100 pairs of castings to support initial production. Three weeks after this request, the parts were delivered.

IMI Rapid Prototyping was then invited to tender for the production quantity of 1000 units. We had already built the SLA models and tools, and costs for these had been covered in the original contract. Following discussions with the foundry, we agreed that if we could amend the tool slightly to incorporate the in-gate into the waxes as supplied, then significant cost reductions could be applied. These costs were passed on to our customer, who subsequently placed the order on us and gave us a delivery schedule of 125 sets of castings per month.

All waxes were produced from the original set of tooling and subsequent to production completion, a further 100 sets of waxes have been produced in-house for samples and exhibitions. These tools have therefore produced in excess of 1250 sets of wax patterns and remain as good today as they were at the outset.

4   CASE STUDY – ELECTRODRIVES MOTOR HOUSING

Electrodrives Limited were requested to produce eight complete sets of motor housings for a bench-mounted vacuum pump. The housings were to be used to test the overall performance of the pump prior to production. Material specification was aluminium and a cost comparison had taken place between Quick-cast SLA models, waxes via our own methods, or machining from solid. The components comprised a main body of 280 mm diameter by 240 mm deep and an end-cap 280 mm diameter by 60 mm deep.

The outside of the housing comprised several fins to act as heat-sinks and the inner was to be machined after casting to ensure clearances between the motor and the housing were correct. There was also a centre bore for the main pump shaft. In terms of cost, our method and the machining from solid route were very similar for eight off, however, if any further sets were required, then machining from solid became uneconomical.

SLA models were produced over the Christmas shutdown period and taken to the client for approval on 3 January. Once approved, silicon tools were made. Due to the design of the component, a six-part tool was needed for the main body to allow splitting along the lines of the fins. Ten sets of wax patterns were produced and shipped to the foundry for casting. At the request of the foundry, we also cast in-gate waxes to their design from a separate tool and supplied these with the component waxes for attaching at the foundry.

When cast, the parts needed heat-treatment to TF condition and this was done at a local sub-contract heat-treatment company in one furnace load. The parts were then supplied to Electrodrives un-machined since they had undertaken to do the machining themselves.

After assembly and testing, two major problems were found. Firstly, the clearances required and machining allowances made for the motor were incorrect and it was therefore not possible to run the motor for long periods. Secondly, there was a severe resonance problem due to the design of the chamber. These problems could not be overcome easily and required some re-design work. Electrodrives were faced with the cost of new SLS models and tooling and also the delays involved in getting parts of the new design.

After detailed discussions on their requirements, we decided that we would carry out the amendments needed to the existing SLA models. These models had survived the silicon tooling process, and were available to aid the discussions. In our workshop, additional struts were added to the main body SLA model to overcome the resonance. These struts were machined from plastic and wood. In order to provide additional machining allowance on the inner bore of the main body, we were able to apply a 3 mm layer of sheet-wax to the SLA model. When all these amendments were completed, we created another silicon tool and started wax production immediately.

By making these amendments by hand, we were able to save the customer the additional cost of a new set of SLA models, in this case, over £8 000. As importantly, the delay in producing new tools was only 24 hours and not 4–5 days as would have been the case with new SLA models.

A further eight sets of castings were then produced, successfully machined, and tested, and another order for 30 more sets was received and supplied from the same silicon tools. Time-scales on the additional sets was not of great importance and these were supplied five weeks after order placement.

5   CONCLUSION

It is hoped that the examples shown above give some idea and feel for the route to metal that we at IMI Rapid Prototyping are able to offer to our clients. We do stress that it is not our claim that this is the only way, or even necessarily the correct way to get to metal castings. We state simply that it is an option we feel should be considered when looking for metal components in prototype or low-volume production quantities.

Rapid prototyping in pattern making and foundry applications

Peter Harrison
W.C.M. Patternmaking Company, Leamington Spa, Warwickshire, UK

1   ABSTRACT

The adoption of rapid prototyping (RP) techniques within the pattern making and foundry industry has necessitated a complete overhaul of skills requirements within the industry. The ever increasing, and soon to be universal, use of CAD by designers, even at prototype stage, has given the production of RP models a distinct advantage over conventional techniques. CAD data can be utilized directly by RP machines to generate a part, thus ensuring an exact replica, without the errors introduced by manual interpretation of drawings, is produced. This revolution has certainly cost jobs and is permanently eroding the skill base of pattern and model making, but only those who can adapt to the new methods of working will survive. Flexibility to give short lead times from CAD model to finished part are the main criteria for success and the prototype pattern maker, with adaptation, can play a vital role in this process.

2   CAD GENERATION

2.1   STL
In the near future CAD models will be almost universally accepted as the only way of transferring data to the point of manufacture from the designer. Unfortunately, Some CAD systems are better than others for the creation of STL files (the type of data used for RP) from the native format. STL files can be prone to problems such as missing or reversed surfaces and often these problems have to be referred back to the customer. The alternative is for the patternmaker to become proficient in CAD, however this is a difficult decision to make. There is the choice of investing many thousands of pounds on CAD hardware, software and training to offer this service or to subcontract this work to a third party.

2.2   CAD design
When a CAD model is built on an RP machine it generates exactly what is drawn. This sometimes has drawbacks. Generally, designers have a limited knowledge of foundry

requirements and consequently many CAD models are supplied with no draft angles, fillets, and machining allowances. These are features that traditionally the pattern maker has added to the basic component design by hand. These requirements need consideration at the design stage to ensure that the foundry can generate suitable pattern equipment from the data supplied. The addition of tapers, fillets, and machining stock to RP models by hand, after manufacture, is possible but this is time consuming and costly – thus undermining the benefits of the RP route. Moreover, the modification of the product by hand invalidates the CAD model and can lead to problems in the future. Considerable time and cost saving can be made when the designer, foundry, and pattern maker work closely together to ensure that the product design incorporates all of the features required for manufacture. Unfortunately, due to pressure of work, it is rare that this close co-operation occurs.

3 RP TECHNIQUES

Once the design has been finalized and the number of castings required has been identified, the process (or processes) for manufacturing the patterns must be decided. As a company we have used nearly all of the RP routes available and there are no firm rules for choosing the material or process for the model. The following sections give a brief overview of the advantages and disadvantages of the different RP processes with respect to pattern making.

3.1 Laminated object manufacturing (LOM)
LOM is used mainly for larger, bulky parts because the paper that it uses is relatively inexpensive compared with resins and powders used in other RP machines. Parts can be built quickly (500 mm diameter x 55 mm deep wheel within a day). The large build envelope of this process means that there are no associated problems with manually joining parts for most automotive applications. Smaller parts can also be generated in LOM and the process produces parts that are fairly easy to finish. The models can be adjusted and modified using standard pattern making machinery and tools. The main disadvantage is the stability of the parts in humid conditions. Figure 1 shows a typical LOM part.

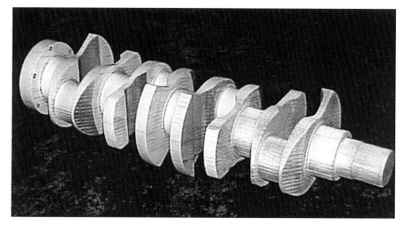

Fig. 1 A LOM model of an engine crank

When parts are manufactured they are contracted by 2–3 per cent in the Z (build) direction. Over a period of time, after manufacture, parts grow into tolerance. Parts may also suffer from de-lamination if subjected to prolonged periods at elevated temperatures and can become unsuitable for casting. However, with the right controls a LOM pattern can be used to produce up to 20 castings, depending on the geometry.

A limitation of the LOM process, particularly in areas with thin wall sections, is the material's susceptibility to 'Air Set' sand formulations, resulting in de-lamination of the pattern. However, heavier section patterns can be used for pre-production runs. Where higher numbers of castings are required it is relatively straightforward to 'clone' polyurethane patterns from the LOM models, which are used for low-volume production. Indeed, often cast resin production equipment is cast from LOM patterns.

3.2 Stereolithography (SLA)

In general, SLA patterns have a longer life than the equivalent produced using the LOM process. SLA parts can be easily finished and any small build steps can be filled and rubbed down to create a suitable pattern for foundry use. SLA parts are fairly stable, usually remaining in tolerance, even in the extreme conditions of a foundry. Recent advances in SLA resins have resulted in a range of more flexible and durable materials, enabling heavily ribbed patterns to be removed from hard sand moulds.

Where parts have to be built in sections they can be joined easily using SLA resins. Moreover, these bonds are comparatively strong and are able to withstand the rigours of foundry use. Larger, bulky, parts are not generally built using SLA because of the high cost, longer lead time and problems with potential distortion. As with LOM, it is possible to cast more durable polyurethane patterns from SLA models.

3.3 Thermojet

Wax parts produced in this system can be used as sacrificial patterns for investment casting. The main advantage is in the production of relatively complex castings without the need for tooling. The use of this technique is increasing rapidly as it offers a cost-effective way of creating complex metal parts directly from CAD models in a relatively short period of time. The wax patterns need to be finished to a high standard because investment casting gives a faithful reproduction of the pattern in metal. Unfortunately, one problem with the Thermojet process is the support system used, which leaves undulations on all downward facing surfaces of the pattern. The supports have to be removed and the surfaces cleaned by hand, which requires some of the skills of a pattern maker to ensure that the models remain accurate and intact.

This process is best suited to small numbers of complex parts that would otherwise require a significant amount of coring to accommodate undercut features. A major advantage with this approach is the designer does not need to add draft angles to the geometry, although radii and machining allowance may be required.

3.4 Laser sintering (LS)

3.4.1 Sintered nylon

Sintered nylon parts can be used as direct patterns for sand casting. They may be more durable than SLA patterns but require more hand finishing and need to be sealed before use (see Fig. 2).

Fig. 2 LS model prior to the production of pattern equipment

3.4.2 Sintered sand
The technique of sintering sand to create cores and moulds is being gradually accepted within the foundry industry as a quick and reliable route of producing prototype castings. As with investment casting the designer does not need to incorporate draft into the CAD model. Complex castings can be achieved without the need for tooling or patterns and excellent location of cores (internal shapes) are virtually guaranteed. In our experience the use of moulds and cores in sintered sand causes no great problems in foundry work and can greatly reduce lead times in many instances.

4   CASE STUDIES

We have been incorporating RP techniques in the pattern making process for the last six or seven years, gradually increasing the percentage of processes and parts each year. In 1999–2000 approximately 75 per cent of all our projects included some form of rapid prototyping. Many projects included two or even three different RP materials to achieve customer requirements of price and delivery. Outlined below are some case studies. Unfortunately, in some cases for confidentiality reasons, they are not all fully detailed.

4.1 Pneumatic tool handle
Norbar Torque Tools Limited manufactures a wide range of hand tools. Several of its customers had asked for a new compact nutrunner. Prototypes were required to ensure that the product met customer requirements before investing in production tooling. The initial requirement was for five nutrunners for in-house testing at Norbar.

4.1.1 Pneumatic handles – five off
The first five castings (pneumatic handles), including heat treatment, were required by Norbar in just 20 days. The casting required the production of a very complex core, which precluded the use of investment casting. The design of the core included areas of 6 mm diameter compound angle holes running through the part (see Fig. 3). Because of the tight time-scales and the complexity of the cores, it was decided to laser sinter them in sand.

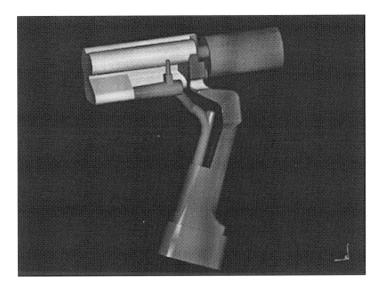

Fig. 3 Norbar nutrunner core(to view this figure in colour, go to colour section)

To produce the sand mould it was decided to produce a LOM pattern, as this was an inexpensive way of achieving a solution within the tolerances and time-scales required. Some manipulation of the STL file was required to add core prints, which locate the core in the mould prior to casting. The external form of the casting had areas of undercuts and it was necessary to take core models off these areas and create core boxes. A reversal was taken off the LOM pattern in polyurethane resin and this was used to check the size of the laser sintered sand cores. Resin patterns were 'cloned' from the reversals and these, together with the runner system, were set into frames for production of the Air Set sand moulds. The pattern equipment and laser sintered cores were supplied to foundry in the confidence that all of the laser sintered sand cores would fit perfectly in the Air Set moulds and casting tolerances would be achieved. The parts were cast, heat-treated, shot blast, and delivered to the customer on time.

4.1.2   *Pneumatic handle – 30 off*

Following testing of the first five prototypes, Norbar decided to make some minor modifications to the design. A further 30 off castings were subsequently required for field trial with a range of customers. As with the prototypes, the new design required a complex core. To laser sinter the cores in sand would have been very expensive but it was just possible to manufacture core boxes within the time-scale (25 days) for production of the 30 castings. Again a LOM pattern was produced of the external shape of the casting, together with a SLA model of the core, including print extensions. SLA was selected rather than LOM because of its better dimensional stability and because the geometry of the core included some fine features.

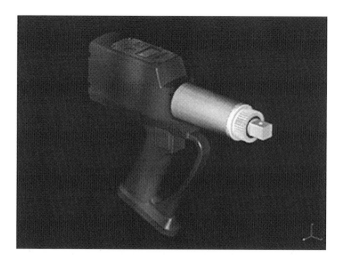

Fig. 4 The finished nutrunner (to view this figure in colour, go to colour section)
(for additional information see www.norbar.com)

The SLA model of the core was fitted into a polyurethane mould taken from the pattern to act as a jig and split into three separate parts that would mould easily. These were used to cast polyurethane core boxes, complete with vents to enable sand to be blown into them. Hand ramming the sand would have been difficult for this geometry and would not have given the required surface finish. Resin patterns to produce AirSet sand moulds were manufactured and set-up for production. The 30 castings required were completed on time.

The 30 prototypes (see Fig. 4) have been evaluated by customers and, based on this valuable feedback, Norbar have refined the design for a third time and the product is due to be launched in the near future.

4.2   Impeller

4.2.1   Impeller – one off
An aluminium bronze impeller, some 300 mm in diameter, was required by a customer within eight working days from receipt of an STL file. The part was relatively complex and using a traditional sand casting route internal and external cores beneath the fins would be required. To achieve the very tight time-scales the laser sintered sand route was the only feasible option. This route carries a degree of risk, as just one sand mould could be manufactured in the allotted time giving one chance of producing a metal part. The STL file was reversed to generate the mould cavity and split into upper and lower plates, complete with location for the internal core and fin sections. The various components of the mould were laser sintered, prepared, and delivered to the foundry who assembled the mould within traditional moulding boxes. Additional sand was rammed around the laser-sintered mould to give an area for the runner system, which was cut in by hand. Metal was cast and the part produced was fettled/shot blasted and supplied to the customer in seven days (five days for file preparation and mould manufacture and two days for casting).

4.2.2   Impeller – three off

Testing of the impeller resulted in major design modifications and additional prototypes were required. Time-scales for delivery of three new castings was just 20 days. There were a number of options for the creation of the new design but it was felt that utilizing LOM and traditional pattern making would be the most cost-effective route and also offer repeatability if required. A LOM pattern was generated, with core prints for the internal feature, together with a model of the core. These parts were hand finished to a high standard. The model of the core was used to manufacture a polyurethane core box. Further core models were moulded where there were undercut areas beneath the fins. These models were used to generate additional core boxes before being assembled on to the main LOM model of the impeller. This pattern was then mounted on an oddside (moulding board) and delivered to the foundry who used traditional moulding boxes and techniques to manufacture three impellers in aluminium bronze.

4.3   Balusters

Corvane specialize in the creation and production of three-dimensional design. Corvane required 200 left- and right-hand brass balusters, to be used in a restaurant to form a balustrade. Time-scales for the production of the patterns was eight days. The parts (LH & RH) were over a metre in length (see Fig. 5) and would not fit on to any RP machine bed without being cut in two. The shape of the parts and data available would make manufacture by hand very difficult so it was decided to cut the parts in two with a positive location and build them by LOM.

Fig. 5 Corvane baluster design

All of the LOM parts were built at the same time and assembled to form the two balusters using simple jigs. Because 200 castings were required it was decided to generate polyurethane pattern equipment from the LOM models. The patterns were mounted on foundry boards (see Fig. 6) with location for conventional moulding boxes for green sand moulding. The sand was rammed around the pattern, which was then removed, leaving a cavity into which the metal was poured. The parts were cast, fettled, brushed, finished with an acid patina, and installed into the restaurant on time (see Fig. 7).

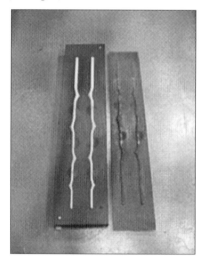

Fig. 6 The patterns produced from the LOM models

Fig. 7 Corvane balustrade installed in restaurant
(for additional information see www.corvane.com)

5   CONCLUSIONS

The incorporation of RP techniques is a natural development within the foundry industry and enables the production of rapid metal parts from rapid models. If metal parts are to be manufactured then the processes to be used need to be fully understood by all parties involved. A close liaison between designer, pattern maker, and foundry from the outset of a project is vital to ensure that a project runs smoothly and time-scales are met.

RP is a growing and ever-changing industry and future developments will inevitably reduce the cost and time for manufacturing models. The question for the prototype pattern maker is where will it eventually lead? The ideal way of creating metal parts is directly from CAD data without the need for models or patterns. The direct production of metal parts by laser sintering, for example, is now possible, albeit it limited in terms of part size and materials. When these processes become quicker, cheaper, and totally reliable for all metals, perhaps the prototype pattern maker might have to broaden his skills again.

ACKNOWLEDGEMENTS

The author would like to thank the companies Norbar Torque Tools Limited and Corvane for permission to use the case studies included in this paper.

Rapid prototyping – enhancing product development at Parker Hannifin

Anders Karlsson
Research and Development Department, Parker Hannifin AB, Mobile Controls Division, Borås, Sweden

ABSTRACT

This paper covers the use of laser sintering (LS), both the successes and problems encountered during some projects carried out at Parker Hannifin, Mobile Controls Division in Sweden.

Rapid prototyping has been a part of projects since 1996 and used on a broad variety of designs. It has mainly been used to produce sand cores for hydraulic castings, to date about 150 castings have been produced. Most of them successfully, but some designs have encountered problems.

Rapid prototyping shortens the lead times in projects with parts requiring pattern equipment. It can cut down the overall project cost for small series production and prototypes when compared to the use of traditional pattern equipment. There is also the added benefit of allowing design changes to be incorporated without the additional costs of making or modifying pattern equipment.

The disadvantages of rapid prototyping for hydraulic castings can be summarized as inherent casting problems such as porosity, core distortion, and poor surface finish causing lower fatigue strength. Porosity and core distortion cannot be related to the prototyping process alone. These kinds of problems are attributable to 'first-off' parts off serial equipment using traditional methods.

The benefits have so far well exceeded the disadvantages and rapid prototyping will continue to be the fundamental of prototyping at Parker Hannifin.

1   THE COMPANY

Parker Hannifin is an American corporation and a leading provider of motion and control technologies. Solutions are provided for a wide range of markets such as aerospace, automation, climate and industrial control, filtration, fluid connectors, hydraulics, instrumentation, and sealing. The Mobile Controls Division is located in Sweden and England and specializes in mobile hydraulic systems. We develop and manufacturing hydraulic valves, pumps, motors, and electronic control systems. Figure 1 shows a typical product, P70CF, a directional control valve.

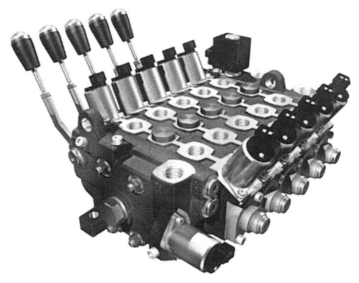

Fig. 1 P70CF

Except for machining and assembly, the company does not have any in-house equipment for prototyping. Services and components, mainly hydraulic castings produced via rapid prototyping, are purchased from various sources all over Europe.

2   BACKGROUND

The demand to shorten lead times in product development projects has been increasing over the last decade, mainly due to changes in market forces and the requirement to become more dynamic. In order to meet these demands, a new model for running development projects was adapted in 1995. During work with that model several methods and technologies, previously unused, were incorporated in the process. Two of those were the use of 3D CAD (solid modelling) and rapid prototyping.

Since then rapid prototyping has been an essential part of the development processes. Several different methods have been used to produce a wide range of prototypes. LOM is used to produce complete hydraulic valves for visualization and packaging purposes. SLA is used mainly as a step to make silicone moulds for producing resin parts, such as covers for electronic units, and parts with very complex geometry, such as control joysticks. The most commonly used method, though, is laser sintering (LS) – in this case, of sand. Another name for this process is direct cronin. This has been used for production of prototype hydraulic valve bodies.

The requests for physical prototypes remain high even though most properties can be simulated and analysed by a range of different methods. Modern CAD systems can easily be used for verifying geometric constraints and to calculate mass properties. Finite element analysis can be used for mechanical properties and fatigue analysis; computational fluid dynamics (CFD) can be used to optimize flow paths and other fluid-related properties. Software packages are also available for calculation and optimization of both static and dynamic valve characteristics. Nobody, however, can simulate the sense of how a complete hydraulic system, for example on a crane, will respond dynamically when it is operated. For this reason prototypes are necessary.

3   ADVANTAGES OF SAND SINTERING

3.1   Methods and cost

There are different methods that can be used to produce prototype castings. The complete mould can be sintered in two or more separate parts, depending on complexity. It is also possible to sinter the cores together with the shells, the cores being well supported by the outer mould shell, allowing very complex designs to be produced easily. This alleviates any effort in designing separate cores and determining how they may be located in a mould. It might be necessary to split the mould into several parts, however, if the design is too complex, since it may be difficult to remove unsintered sand. Using this method no 'draft' or 'clear line of draw' is required for assembly of a mould. This can save a lot of time but also permits designs that are impossible to produce in serial production equipment.

Another hybrid route is to sinter only the cores and use traditional prototype pattern equipment for the shells (cope and drag). This requires a closer co-operation between the different parties, designers–foundry–RP bureau–pattern makers, in order to obtain a core that fits into the mould.

It is generally considered that the break-even point for sand sintered moulds compared to traditional machined pattern equipment is in the order of 1–10 castings – but this depends on the geometry. For the hybrid route, figures vary up to as much as one hundred castings. Thereafter, traditional routes are more economical.

The first project using LS was to develop a very complex casting, dedicated to the control of a dumper. The core was built in seven parts, glued together, and successfully cast. All five sets delivered castings, which were usable as prototypes after machining. Figure 2 shows the core design.

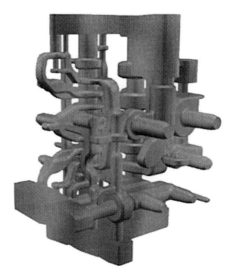

Fig. 2 Core from steering and dump valve

3.2   Lead times

Lead times for producing prototypes are dramatically reduced. Using solid modelling combined with laser sintering of sand cores and production of resin prototype patterns has reduced lead times from 3–6 months to 2–4 weeks after submission of suitable files. This is compared to the days when 2D drafting and traditional equipment were used.

Using the method of sintering to produce a complete mould can reduce the lead time to a few days. The time saving in producing the casting is not the whole story though. It should also be considered that the overall lead time, including both design and production of the casting, is reduced – since no drawings are necessary.

3.3   Freedom of change

In every development project design changes occur frequently. The reasons are many and usually result in a physical change in the design. Besides the inevitable physical errors that may be made during the design cycle of any new product, legislation governing the use of a product or the conditions in which it was originally intended may alter. This is particularly true of projects whose product development lead time may be as much as a year.

Altering the design early in the development cycle does not usually have much effect on the overall development costs. However, if changes occur late in the design cycle the cost is usually very high. It is therefore essential that changes are made as early as possible.

The use of rapid prototyping provides a great opportunity to test different and unconventional solutions before deciding upon the final design. Doing this without having to invest in unnecessary pattern equipment cuts the overall project cost.

During one of Parker Hannifin's projects, the company was forced to make some dramatic changes due to late customer requests, as well as new law requirements regarding safety functions. Cost estimates for these major changes would have resulted in expenditure in the order of $150 000 if the first castings had been produced with traditional pattern equipment and methods.

The ability to produce castings so early in the project not only gave the production engineers an opportunity to optimize NC-paths and cutting data, but also to verify fixtures and pick-up points. In addition, they were able to suggest minor design changes in order to minimize future problems such as deburring and cleaning. Previously, changes like these would not have been considered worthwhile due to the cost of changing pattern equipment.

3.4   Verification of functional requirements

A prototype casting produced using rapid prototyping can be used for almost any kind of verification test. Because the material and shape are the same as though produced from serial production equipment, it is possible to verify machining programs, pressure drop, and other hydraulic characteristics.

One point to consider with respect to the shape is that when producing serial equipment some compensation is made for wear and surface coating, for example. This means that the geometry or the flow paths, in this case, will be slightly smaller compared with the prototypes – having more or less nominal geometry. This has given better results from the prototypes when compared to the first samples off the serial pattern equipment, especially when measuring pressure drop.

Since prototypes of this kind are fully functional, it has been possible to use them for early verification of fatigue and endurance tests. The results show that the prototype castings produced from the laser-sintered cores show a lower fatigue strength compared to a casting produced from a hot-box core (serial production core boxes). Comparing FE analysis and material specifications with endurance tests has shown that the fatigue limit is reduced by approximately 50 per cent. This difference can be explained by the comparatively rough surface left on the surface of a casting from a laser-sintered core as opposed to a shell-moulded hot-box core. Since these tests were made on castings produced in grey iron further work is required to verify these figures because of its relatively large tolerance on yield stress.

3.5   Post-production engineering

Experience gained from the prototype phase has helped production engineers avoid problems normally encountered when traditionally proving out series pattern equipment. Since the casting technique used for both prototype and serial production is similar, experience gained is used to help solve gating and feeding issues.

This experience also gives a good picture of how the design is suited for casting. One such prototype design continually produced scrap prototype castings. One feature of the core was bending during casting and resulted in no machining allowance. After a redesign, which involved adding additional support to the core, no further problems were encountered. As a consequence none of the serial production castings from this design have suffered from this problem.

One further unexpected consequence of this route has been its effect on initial samples from serial production pattern equipment. All castings to date have been delivered within dimensional specification. Normally the inspection reports in some instances run to a few pages but not any longer.

Some of the castings were given to another foundry, not involved in the prototype process, for serial production. Using another casting method, the experiences gained from the prototyping process were unusable for them and those castings did not meet the specifications as well as the others.

4   QUALITY OF PROTOTYPE CASTINGS

4.1   Accuracy

The first sample castings made from laser-sintered cores were measured and evaluated quite thoroughly. The results were not perfect but better than expected; most of the records were within specification, grade CT8 according to ISO 8064-1984.

Having a good understanding of how the cores are manufactured using this technology has given us a better understanding of why inaccuracy can occur. After a core has been laser sintered, excess sand has to be carefully removed from around the core. The core must then be surface treated with a micro-blowtorch or hot air gun. This is done in order to give the core a higher degree of surface resistance for the next stage of the process. The core is then put in a box, embedded with glass beads, and heated to 180 °C. As the core is heated, the resin's viscosity falls before fully curing and achieving full strength. If the core is not fully supported by the glass beads, the core will sag under its own weight resulting in distortion. This can also occur if too much heat is applied with the micro-blowtorch or hot air gun while preparing the core's surface.

Another potential problem can occur while heating the core during post cure, when it is important that the core reaches a uniform temperature. Failure to heat the core uniformly may result in certain features remaining partially cured. This can in certain circumstances result in the core distorting during casting as the semi-cured section becomes viscous again. Alternatively, a large sintered core will take many hours for the temperature to soak through the sintering container and the glass beads. This may result in the centre of the core being adequately cured but the extremities of the core being over-sintered. This results in the core becoming very fragile as the resin becomes over cured and loses its strength.

Some of the inaccuracy can occur with the CAD system. When a model is complete it has to be exported into a neutral format, STL. This is a triangulated format where the user has the ability to specify the size of the triangles. Specifying a small size gives good accuracy but results in very large file sizes. There is also the risk that too large a triangle size is used resulting in reduced accuracy and loss of definition.

4.2   Surface finish

A core from a rapid prototyping system has a rougher surface than a serial produced shell core. This is due to some grains of sand at the outer boundaries of the core being sintered to the main body while others remain unsintered. With traditional patterns and core boxes, all sand grains are aligned to the surface of the tool giving a smoother shape. Where a better surface finish is required a coating can be used. The normal coatings used in serial production contain solvents. These solvents have to be dissolved before casting. A normal procedure to do this is to put the cores in an oven and burn out the residual solvents. However, this may affect the cure of the core resulting in over sintering or at worse potential loss. It is therefore recommended that the coating be applied before it is post sintered.

Even if the core is treated in order to remove the solvents, the coating itself generates gas when in contact with the molten metal in the mould. This may cause problems in the casting process. It is therefore advised that use of coatings be considered carefully since the benefits of achieving a better surface finish may be compounded by additional problems with both the core and casting.

4.3   Porosity

The resin content of the sand used in the laser sintering systems is high (5 per cent) by comparison to the zircon sand used for series production shell moulding. This results in a lot of gas evolving in the mould during casting. This is particularly true when casting hydraulic valves, since the cores are almost completely enclosed by metal. It has therefore been essential to incorporate as much venting as possible in the mould and core to allow the gas to escape. Failure to provide venting will result in the castings suffering from porosity and they will be unusable for further processing. Figure 3 shows a section of a hydraulic valve casting, suffering from porosity caused by both shrinkage and captured gas.

Fig. 3 Section of hydraulic valve casting

The core was also subject to distortion caused by the heat. A computer casting simulation showed that this was the hottest part of the casting with the slowest solidification rate. Altering the feeding system combined with better venting resolved the problems in this area. Venting can either be built into a core by the laser sintering process or be drilled after being post cured.

The addition of venting into a complex core that is to be laser sintered can be relatively risky, especially for the types of geometry that are of interest to Parker Hannifin. The vent holes can make a fragile core even weaker, making removal of unsintered sand and handling a very delicate process. In these instances, it may be easier to drill this type of core after being post sintered, as the cores are usually much stronger. Even so, care has to be taken with post-sintered cores, as they can still be easily broken.

The geometry of the core shown in Fig. 4 caused a lot of problems in both handling and venting it. The core prints were initially large and heavy, and were carried by comparatively small and weak features. Some of the cores broke during transportation and others during casting. The core prints were subsequently redesigned as a frame to give additional support to the core. Additional features were also added and modified to increase the strength of areas prone to breakage during handling and the post sintering cycle. The core, as can be seen, was originally produced in one piece. It was decided to split the design for increased strength and to help incorporate venting. The resulting core was successfully built with venting and subsequently glued together prior to casting. The action resolved the problems and resulted in a better casting. Figures 4 and 5 show the old and new core designs.

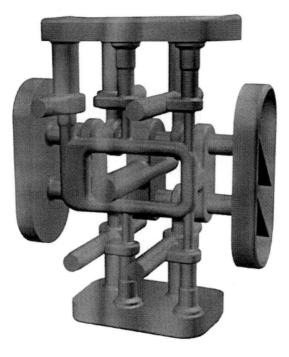

Fig. 4 Old core design

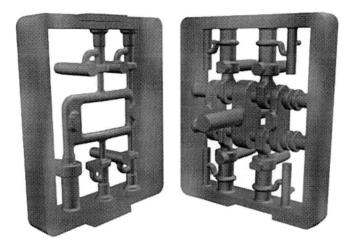

Fig. 5 New core design

Cast tooling with rapid prototype patterns

Greg Redden
CSIRO Manufacturing Science and Technology, Preston, Victoria, Australia

ABSTRACT

The most commonly used methods for making prototype diecast parts, such as sand, plaster, and investment casting, do not produce parts with similar mechanical properties to those found in production diecastings. Similarly, although to a lesser extent, this is also true for plastic injection moulded parts. The only way to ensure prototypes have identical properties to the production part is to produce such parts using the normal production process. This requires the use of steel dies. However, such steel dies are expensive and require substantial lead times, making them unsuitable for prototyping or short run production purposes.

With the advent of rapid prototyping and CNC machining, complex plastic patterns can now be produced in a matter of hours. This has opened an opportunity for precision ferrous casting processes for manufacturing prototype and short-run tools using rapid prototype patterns.

1 INTRODUCTION

The manufacture of prototype parts for design evaluation is becoming increasingly common. The most common way to make prototypes of diecastings is by sand, plaster, or investment casting and direct CNC machining from solid stock. Spin casting is also popular for small components in low melting temperature zinc-base alloys. Commonly, the patterns for these casting processes are created using rapid prototyping methods such as stereolithography, selective laser sintering, or laminated object manufacture. CNC machining of polymers can also be used.

Sand, plaster, and investment casting of parts are suitable for early design testing. However, the mechanical properties are not the same as those of a real diecast part. This is mainly due to differences in the cooling rates of castings made in sand, plaster, or ceramic as compared to those obtained in a steel diecasting die. For many applications, these prototype cast parts can be used for mechanical testing with appropriate design factors applied to compensate for the differences in properties. Typically, 70 to 80 per cent of the yield strength found in a real diecasting can be expected (1, 2). However, there are few data published on fatigue and impact test results from which to derive compensation factors.

Should the application for which the diecast component is designed require stringent mechanical or functional testing, this leaves the only viable option to be the manufacture of a steel tool with which prototype diecastings can be manufactured using production equipment. Tools made for prototyping purposes also have the potential to be used for production purposes. Unfortunately, even prototype steel tools produced using conventional machining techniques are expensive and have long lead times. Less expensive and, more importantly, faster methods of producing production tools need to be developed.

Cast net-shape tooling has been employed for at least the past forty years for the manufacture of tools and dies for a variety of applications such as forging, pressing, and diecasting as well as glass, rubber and plastics moulding. Prior to the advent of rapid prototyping, these cast tools were produced with conventionally manufactured patterns. Such patterns take considerable time and skill to prepare, especially if complex shapes are involved. However, in recent times the ability to rapidly generate patterns using rapid prototyping has seen resurgence in interest in cast tooling.

2   THE PROCESS

The process usually begins with a CAD model of the part from which a pattern is produced by a rapid prototyping technique. If gate and runner geometry cannot be easily machined into the final casting, these features are best defined and CAD modelled at this time. As this process uses a recoverable pattern, the pattern must be removed from the mould after moulding. If the geometry of the tool incorporates deep narrow slots and features with little or no draft then removal of the pattern will be very difficult. In these instances a slightly flexible pattern will greatly increase the chances of successfully removing the pattern.

Rigid patterns can be made using any of the commonly available rapid prototyping processes such as stereolithography, laminated object manufacturing, or selective laser sintering. Patterns can also be made by the CNC machining of polymers or by traditional techniques. The optimum pattern manufacturing method is best decided on a case-by-case basis. The manufacture of rigid patterns requires the tools to be modelled using CAD, starting with the part geometry. Figure 1 shows CAD data of half a die-casting tool that was modelled from part geometry. If solid models are used then is it a relatively straightforward task to subtract the part from the die-block, resulting in the mould cavity.

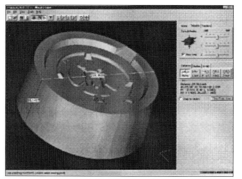

Fig. 1 CAD model of tool

The procedure for making flexible patterns is best suited to parts that allow straight parting lines. The procedure begins by building models of each section of the part, as defined by the parting lines. These define the sections of the part and thus the shape of the tool cavities. These sections of the part are built as rigid models with small allowances added for metal contraction and finish machining to ensure proper shut-off on parting surfaces. The models are then attached to the bottom of separate moulding boxes. The boxes are the size of the flexible tool patterns to be produced. The moulding boxes are then filled with a flexible polymer, such as polyurethane, resulting in flexible patterns of the tool sections.

The use of flexible patterns has the disadvantage that dimensional accuracy is compromised as they can be distorted in the moulding process. Every additional moulding step in the process also introduces errors and therefore flexible patterns are only used when rigid patterns are considered unsuitable.

Figure 2 shows a rigid pattern consisting of a Selective Laser Sintered glass filled nylon prototype, for the complex geometry, with traditionally made patterns for simple geometry such as the runner and gate.

Fig. 2 SLS Pattern on moulding board

From the tool patterns, precision ceramic moulds are made using a variation of the Shaw Process (3). The Shaw Process uses a ceramic that is poured over each pattern as slurry. The slurry accurately reproduces the finest detail of the pattern. After a few minutes, chemical action sets the ceramic slurry and the pattern is then removed from the mould. Before casting, the moulds are fired at high temperature in a furnace. The moulds are then assembled with relatively inexpensive sodium-silicate bonded, silica sand lids. As the back faces of the tools do not have demanding dimensional or surface finish requirements, there is no significant benefit from using the more expensive ceramic material for these lids. After assembly of the moulds, the tools are cast in the desired grade of steel.

Assembled moulds being cast are shown in Fig. 3 and a casting with the gate and runner still attached is shown in Fig. 4.

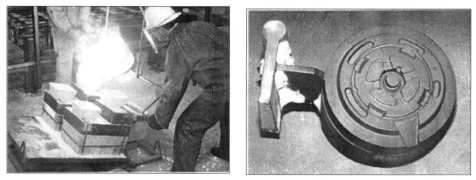

Fig. 3 Casting of tool inserts   **Fig. 4 Unfinished casting**

Before the tools can be used, finish machining is generally necessary on the backs and sides of the tool and, if necessary and practical, on the parting faces. This machining is, typically, simple turning or milling. Features such as ejector pin holes can also be bored at this time.

The final result is a tool set, such as that shown in Fig. 5, that could take as little as two weeks to produce from CAD data to finished tool set.

Fig. 5 Tool inserts after machining

The process is summarized in flow chart form in Fig. 6.

3   PROCESS LIMITATIONS

This process does have limitations. The most significant is the inability to accurately prepare moulds with which to cast tools that incorporate deep, thin slots. This is important as a large number of diecast parts have such features.

The achievable accuracy also requires further improvement to enable an increase in the number of diecast tooling applications for which this process is suitable.

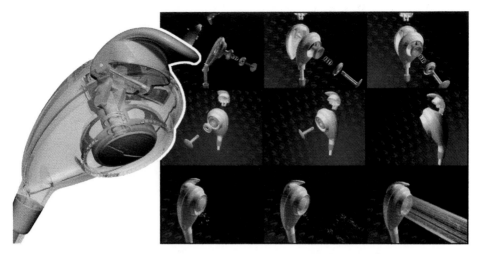

Fig. 4 Computer visualization and stills from a computer animation used to demonstrate the construction and function of a new product concept. The animation was used to sell the concept into key retail outlets to secure the essential routes into the market. 'Freeway' Shampoo Spray for Croydex
(Figure 4 in the paper by Steve May-Russell and Martin Smith Section 1)

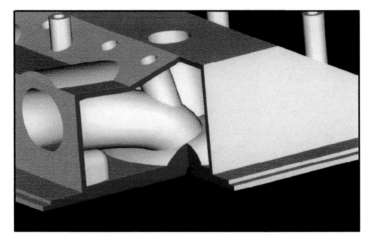

Fig. 6 from the paper by C Driver – Section 1

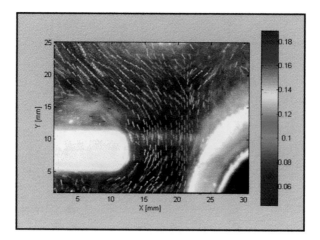

Fig. 10 from the paper by C Driver – Section 1

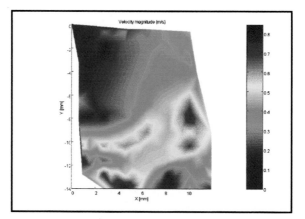

Fig. 11 from the paper by C Driver Section 1

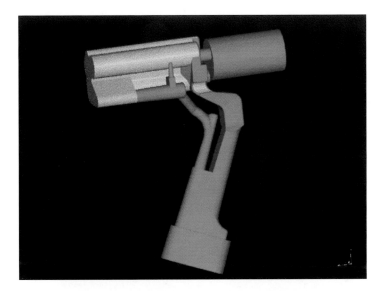

Fig. 3 Norbar nutrunner core (from the paper by Peter Harrison – Section 2)

Fig. 4 The finished nutrunner (from the paper by Peter Harrison – Section 2)

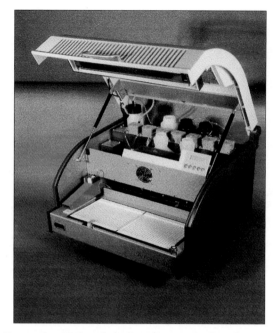

Fig. 3 The complete DNA analyser (from the paper by Tony Sands – Section 4)

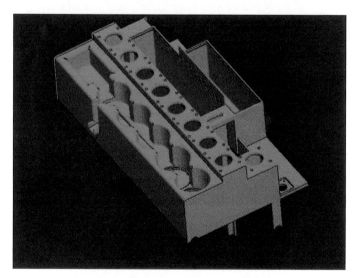

**Fig. 4 The main chassis component in STL format
(from the paper by Tony Sands – Section 4)**

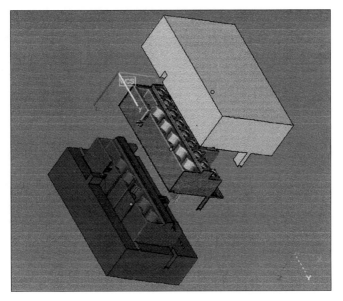

Fig. 5 Virtual tooling created in Magics RP (Materialise)
(from the paper by Tony Sands – Section 4)

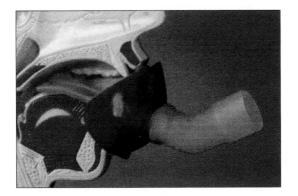

Fig. 4 (a) SLA and vacuum cast assembly of respirator mask for its design evaluation

(from the paper by Joel Segal – Section 4)

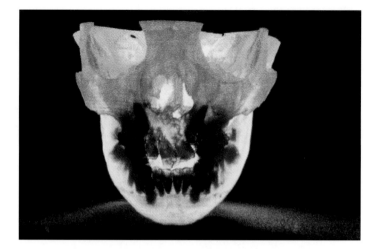

Fig. 1 Skull model with teeth marked in colour; colour stereolithography technique (from the paper by Kris Wouters – Section 5)

Fig. 3 The triangle model of the Cyclone scan (from the paper by Chris Lawrie – Section 5)

**Fig. 5 The ModelMaker system at work
(from the paper by Chris Lawrie – Section 5)**

**Fig. 6 The triangle model of the head
(from the paper by Chris Lawrie – Section 5)**

Screw diameter: 600 µm
Hull diameter: 650 µm
Total length: 4 mm

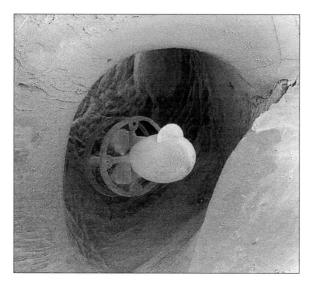

**Fig. 3 Micro submarine CAD image and the micropart seen inside a human artery
(courtesy of Microtec Gmbh, from the paper by William O'Neill – Section 5)**

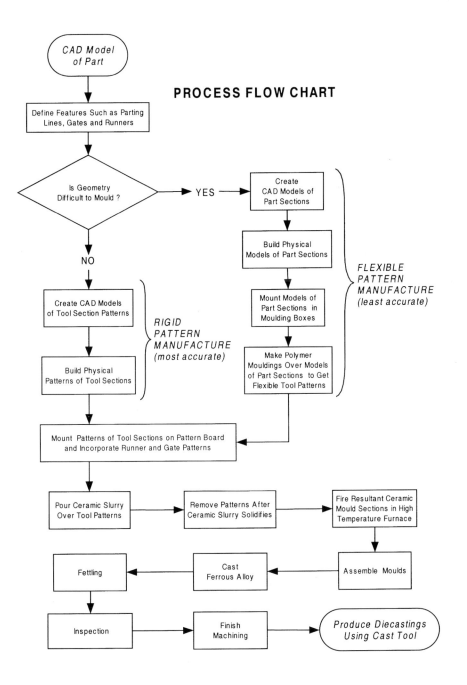

Fig. 6 Flow chart for production of cast tool using ceramic moulding

Both these limitations are directly related to the quality of rapidly generated patterns that can be made from CAD data. In some instances, such as those where the geometry is not of high complexity, it continues to be more economical to use conventional machining than pay the relatively high cost of prototype patterns for use in a casting process that requires some, albeit simple, finish machining operations.

Ideally, the as-cast tolerances of parts produced using cast prototype dies resulting from rapid prototype patterns, should be considerably better than those resulting from methods currently used to prototype diecast parts.

4   CONCLUSION

As less expensive and faster ways of prototyping diecast parts are evaluated and used, it is apparent that every process has limitations. The production of prototype parts by sand, plaster or investment casting does not produce parts that have mechanical properties representative of production diecast parts. Advanced rapid prototyping techniques that employ powder metal processes are relatively unproven and have size limitations. This leaves the options of CNC machining with its usually long lead-time and expense, and casting processes such as that described in this paper.

Rapid prototyping and CNC machining can produce complex plastic patterns in a matter of hours. This has opened an opportunity for precision ferrous casting processes, using these patterns, for the manufacture of prototype and short-run tools.

In some cases the part requirements will be beyond what can be achieved using tooling produced by a casting process due to the cumulative errors in prototyping, moulding and casting. In other instances, the geometry of the tool may better lend itself to high speed CNC machining. However, for some applications, a prototyping and casting approach to produce tooling can realize the following benefits:

- Reduced time to produce prototypes and, with process improvements, a large range of short-run production tooling.

- Reduced cost compared to conventional machining.

- The opportunity to cast cooling passages into the tool exactly where they are most effective. This is of significant potential benefit for tools with complex geometry.

REFERENCES

1.   **Fantetti, N., Pekguleryuz, M.,** and **Avedesian, M.** (1993) Magnesium plaster cast prototypes versus diecastings – a comparative evaluation of properties, Institute of Magnesium Technology, Canada, May.

2.   **Warner, M.** (1996) Rapid Prototyping for Die Casting: Today's Applications and Future Developments, *Diecasting Engineer Magazine*, March/April.

3.   **Shaw, N.** Method of Making Moulds, US Patent 2,795,022

Section 4

Rapid Tooling

Overview of rapid tooling

David Wimpenny

INTRODUCTION

The majority of rapid tooling methods are by no means new, indeed they have coexisted with conventional machined tooling for many years. The re-emergence of these methods in the 1990s was a logic progression from rapid prototyping. Inevitably no sooner had rapid prototyping provided a method of making models or visual prototypes quickly, then there was demand from industry for quicker ways of making technical prototypes, manufactured in the correct material, using the appropriate production method. Rapid prototyping offered a potential solution to this problem, by supplying the master pattern from which rapid tools could be generated and also by the opportunity to manufacture both prototype and production tools directly.

Rapid tooling methods can be divided into two basic methods:

Indirect tooling – tooling generated using a master pattern (often a rapid prototype model but also hand made or machined pattern). Methods include;

- Cast resin
- Cast metal
- Metal spray
- Sintered metal tooling

Direct tooling – tooling manufactured directly on the rapid prototyping machine. Methods include;

- Direct AIM stereolithography tooling
- Laser sintered metal
- Laminated tooling

CAST RESIN TOOLING

Cast resin tools have been used for polymer processing since the 1940s. Despite the recent introduction of a number of new resin systems, specifically formulated for injection moulding tooling, the basic process remains largely unchanged. The resin, typically a metal-filled epoxy resin, is simply cast around a conventional or rapid prototype master pattern, to form the tool cavity. To achieve the desired properties the tools may need to be post-cured at an elevated temperature. For high-pressure moulding processes, such as injection moulding, cast resin tool inserts are usually placed within a steel bolster restrain the clamping/injection forces. Although the process seems very straightforward, the results are highly dependent upon the part geometry and careful control of the toolmaking process.

CAST METAL TOOLS

A number of metal casting techniques can be applied to the manufacture of tooling, using both conventional and rapid prototype patterns.

Sand casting is commonly employed to manufacture near net shape tools, which usually require final machining of the moulding surface. The process is often used for manufacturing relatively large tools in aluminium or cast iron for low-pressure moulding processes. Rubber-plaster casting (and very infrequently investment casting) can be used to produce net shape tools which only require machining of the shutout and sliding insert locations. In rubber-plaster casting a silicone copy of the required tool cavity is cloned from a master pattern. A gypsum material is cast around the silicone. Once this material has solidified the silicone pattern is removed and non-ferrous metal (usually zinc or aluminium) is then cast into the space left by the silicone to produce the tool insert.

METAL SPRAY TOOLING

The basic principle of convention metal spray tooling is that a thin layer (1–2 mm thick) of metal, usually low melting point material, is applied to a master pattern. Then, depending on the particular moulding application, the shell is supported with a suitable backing material, typically a metal-filled epoxy resin. Arc spraying of low melting point zinc alloys is the most frequently employed metal spraying route for the manufacture of tooling. This process has proven to be very effective for manufacturing low-pressure tools for vacuum forming, RTM, and blow moulding. Unfortunately, thermal spraying is a line-of-sight process and this can impose severe limitations for the manufacture of more complex injection moulding tools. To provide a surface coating with acceptable integrity the spray must applied, as far as possible, perpendicular to the surface of the pattern. Unfortunately, component geometry can often preclude this and as the angle of impact tends towards 45 degrees the strength of the coating falls sharply. Applying a coating into thin slots or holes also presents a problem and it is common practice to incorporate solid metal sections to form difficult-to-spray areas of tools.

SINTERED TOOLING (INDIRECT)

There are a number of proprietary techniques (Keltool, NDM, etc.) which allow tools to be constructed by fusing powder (usually in the form of a slurry) around a master pattern. The powder is held by a temporary binder which is subsequently removed and the metal powder is sintered to form a solid tool insert. This method enables relatively hard steel powders to be processed. Moreover, multiple inserts can be manufactured from one master pattern. These techniques can faithfully reproduce the surface finish of the master pattern. The accuracy of this process is claimed to be very good, particularly for small inserts under 100 mm cube. For larger inserts it may be more difficult to hold tight tolerances.

DIRECT AIM TOOLING

Direct AIM stereolithography can be a very rapid and effective method of manufacturing small prototype injection moulding tools. As with the other forms of direct tooling the tool is built directly on the rapid prototyping machine and so a CAD model of the tool is required. This has been a major drawback of direct tooling techniques in the past but new STL file manipulation software can produce simple tool designs very quickly. Despite the low heat deflection temperature of the early resin systems direct AIM tools performed surprisingly well, provided that the temperature of the tools was carefully monitored. New high-temperature resins have now been introduced, specially formulated for the manufacture of injection moulding tools. Although even these new resins cannot match the combination of properties offered by cast resin tooling systems, direct AIM tools are regularly used to produce up to 100 prototype mouldings. Many of the same design criteria that apply to cast resin tooling also apply to direct AIM tools, for example thin up-stand may need to be produced using metal inserts and it is generally recommended to mount the inserts in a steel bolster. The thermal conductivity of direct AIM tools is very poor and consequently prolonged cycle times are required, together with very careful cooling of the tool surface using air blasts. Direct AIM tooling is particularly suited for the manufacture of small tools. As the tools increase in size, the cost and time associated with their manufacture becomes a significant factor.

LASER SINTERED METAL TOOLING

The manufacture of metal tools by laser sintering was hailed as one of the major breakthroughs in rapid prototyping in the mid-1990s. Unfortunately, to date the process has been slow to be adopted commercially.

The two main organizations, which manufacture laser sintering machines, have adopted slightly differing approaches to the sintering of metal parts. DTM Corporation advocate the use of a metal powder coated with a polymeric binder which is fused by a CO_2 laser to form the desired geometry. Once the laser building process has been completed the excess powder must be carefully removed from around the tool insert and it then undergoes debinding, sintering, and infiltration processes to produce the finished tool. EOS adopted a slightly different approach. The direct metal laser sintering (DMLS) process a high and low melting point metal are combined.

The laser melts the low-temperature material, which forms a matrix around the higher temperature material to produce a solid object. Using this process, tooling inserts can be produced directly without the need for a furnace cycle.

Despite the significant improvements in materials for laser sintering, concerns regarding the accuracy of the tooling inserts produced still remain. Until the process can offer precision comparable with conventional tooling methods, the use of the process for tooling will be severely restricted.

LAMINATED TOOLING

The concept of rapid laminated tooling is very similar in many respects to laminated object manufacturing (LOM), however, as opposed to employing paper which is bonded with a thermoplastic binder, laminated tooling is most usually formed from relatively thick steel sheets which are bolted, bonded with adhesive, or brazed together.

Rapid laminating tooling predates rapid prototyping by several years, indeed the first work on laminated tooling by Professor Nakagawa was first published in 1980. With the advent of rapid prototyping a number of other research groups have undertaken studies in this area. Despite the large body of research conducted in this field, the level of commercial exploitation of this process has been surprisingly low to date.

SUMMARY

In this Section, a range of rapid tooling methods will be described in more detail and the potential benefits and problems associated with each approach will be illustrated with the aid of case studies.

Flexible silicone and rigid cast resin tooling are described and compared by Tony Sands of RIM-CAST, who also identifies the potential role that rapid manufacturing may play in eliminating the need for tooling altogether in the future. Joel Segal, Nottingham University, shows the potential cost and time savings which can be gained through the use of cast resin and metal spray tooling methods. In the next paper Dr Trevor Illston and Simon Roberts describe direct AIM tools and selective laser sintered copper polyamide tooling for prototype injection moulding and forming wax patterns for investment casting. Then Jouni Hanninen *et al.* describe laser sintered metal tooling and illustrate the application of this technology in both injection moulding and die casting using industrial case studies. In the final paper in this Section Amanda Threlfall, BAE SYSTEMS, outlines the development of laminated metal tooling and illustrates the benefits and potential shortcomings of this technology using tools manufactured during the research programme.

The tooling methods described in this Section only represent a selection of processes available. Unfortunately, it is impossible to recommend one tooling method for all applications, each route has advantages and disadvantages, depending on the particular application. In addition to comparing the so-called 'rapid' tooling methods with each other, it is vital that more conventional tooling methods, such as CNC machining, should also be considered.

The role of rapid immediate production tooling (IPT) in new product development

Tony Sands
Rim-Cast, Kettering, Northamptonshire, UK

1   ABOUT THE AUTHOR

Tony Sands has been associated with the rapid prototyping and manufacturing (RP&M) industry for 20 years and was a founder committee member of one of the first RP associations to emerge. Currently, as a partner in RIM-CAST, he is classed as a 'practitioner' of the technology in that his day-to-day activity is exclusively RP&M orientated.

In parallel with their manufacturing operation RIM-CAST actively support research into new techniques. They have an involvement, along with other commercial companies, in the 'Design for Rapid Manufacture' project initiated recently by Professor Phil Dickens at De Montford University. RIM-CAST's input relates to the analysis of design and product development processes in the context of RM.

This invited article is written hopefully to interest fellow practitioners who will be able to identify with the situations and processes described, but more importantly for new potential industrial users of RP&M who are seeking to realize the full potential of each technical opportunity, particularly rapid tooling capable of extended production use. RIM-CAST have called this development 'immediate production tooling' (IPT) and it represents their method of achieving optimum time to market targets.

Rapid tooling techniques are developing and improving all the time in the RP industry. The author only claims expertise in the techniques pioneered and perfected by RIM-CAST and therefore has of necessity restricted the content of the article to real life experiences with that company on the basis that you can only talk with authority on your own subject. The case histories outlined are typical of many now completed.

2   RP&M – THE BACKGROUND

First let us add words to the letters RP&M. They refer of course to rapid prototyping and manufacturing in the context of taking a product/idea/component from CAD or paper concept through, in the first instance, to a real, solid shape using techniques such as:

- Stereolithography   SLA
- Laminated object manufacture   LOM
- Fused disposition modelling   FDM
- Laser sintering   LS
- Reverse engineering
- Rapid machining   plus other similar emerging technologies.

Having reached the stage in a project where all, or part, or several parts of a product have been physically created by one or other of the above ingenious techniques, it is usual to apply conventional model-making and finishing skills to dress the part or parts into the form required depending on their intended function.

Then what you arrive at in most cases is a *model* of the product. If this model is to be used later as a *master pattern* for some subsequent tooling process, the cosmetic quality and dimensional accuracies are vitally important and *none* of the existing RP techniques will achieve this as an outright result. The skill of today's pattern maker is paramount in achieving model quality. It is a fact beyond dispute that the best models (or master patterns) are still produced by the top echelon of pattern makers whether or not the starting point is an RP model.

This 'RP', or stereomodelling, part of the industry is steadily developing as new materials and techniques yield parts with wider applications and specific physical properties such that they can be used in subsequent processes as:

(a)   full representation in of the final article for actual end use; or
(b)   use in various ways for evaluation purposes or as patterns; or
(c)   in some cases as tools for other manufacturing processes.

The 'M' or manufacturing part of RP&M has at least two connotations:

1   Using CAD/CAM data and reverse engineering, where applicable, to shape solid materials of all types into an engineered form. These can be as one-off components through to volume manufacture, or as tooling for other processes.
2   In the context of this paper, using any of a wide variety of rapid, often low-cost, tooling techniques such as RIM-casting to create market useable product on short development time-scales.

These two categories share the letter 'M' in that manufacture is the theme, but together they encompass a colourful selection of diverse initiatives all driving towards reducing time-to-market for new products.

3   TOMORROW'S RM

In the case of the research taking place at De Montford University, the exciting concept is that, for certain components, designers can approach a product without preconceived geometry or tooling process limitations, and design *end-user* products to be generated directly from software via rapid manufacturing in *one process step*, true RM indeed.

Many products, however, because of size, volume, or cost will perhaps only be able to benefit from selected parts of the RP&M processes, and for these some form of prototype or production tooling will still be necessary.

Also emerging from the RP&M scene are many specialist companies offering to designers and manufacturers the opportunity to 'out source' the skills they require, together with an established selection of bureaux who have invested heavy capital in the latest techniques, enabling clients to buy time, services, models, and even product, created using skills and equipment which at present only major industrial players can afford to bring in house.

The capital equipment for stereolithography is usually expensive. There are, however, several breeds of machines which address the concept of a relatively lower cost, office-compatible, 3D modeller which can be networked into a design office CAD system and used like a printer to produce concept shapes.

4   HOW FAST IS RAPID?

We have touched on prototyping, manufacturing, and rapid tooling, but let us return to this word 'rapid'. The inference in some interpretations of RP&M is that one should strive to do absolutely everything as fast as possible. This can often lead to counter-productive side effects such as exhaustion, frustration, and the unnecessary expenditure of both energy and money.

Let us put this point of speed into perspective in the particular context of RIM-casting. This increasingly popular technique is interpreted as pouring or injecting two-pack, rapid-curing resins into temporary or permanent tooling made of a variety of materials including silicone, epoxy, machined plastic, or metal, or, in the case of RIM-CAST, specially formulated and durable polyurethane tooling resins:

5   TOOLING

There are general distinctions between the various forms of tooling that can be used for casting products in RIM chemicals and a principal consideration is the nature of the tool surface. (We assume in all the cases below that the starting point is the STL model as described earlier.)

5.1   Rigid tooling

This can take the form of tools made of castable materials such as epoxy or rigid polyurethane formed against the surface of the model, which has been suitably prepared with release agent. Obviously such tooling needs to be formed in two or more parts so that the final tool can be opened and closed at the appropriate split lines. Both the surface and body of the tool are usually cast or formed in one piece.

Alternatively, the half tools can each be machined from solid materials including both plastics and metals, in which case the model itself can remain 'virtual' in the CAD system. Again the tool is normally all made of one material.

Rigid tooling, although reasonably durable (possibly hundreds of parts can be produced), nevertheless has two characteristics that can be considered limitations:

1   The product design needs to carry draft angles on all appropriate surfaces in exactly the same manner as for injection tooling, otherwise the product will not de-mould.
2   Should design changes occur after the tool is completed, the alteration into a rigid tool is not a simple matter and in many cases necessitates a new whole (or half) tool.

5.2   'Soft' tooling

This is generally interpreted as being *silicone* tooling although other flexible materials such as elastomeric polyurethanes and latex can be used. The silicone moulding process is a long-established technique and involves the encapsulation of a real model using silicone rubber (usually under vacuum in a large chamber). Once the material has been cured in an oven, the tool is manually cut into half tool or more segments at pre-selected and identified split lines marked on the model. As with rigid tooling the entire tool is usually one material, i.e. a silicone block.

The model is then removed and product cast into the cavity. Again this casting process is performed under vacuum, hence the popular description of the technique as 'vac-casting'. The pros and cons break down in general terms as follows:

5.2.1   Advantages

Because the process takes place under vacuum it is possible to cast thin and extremely detailed sections giving a true representation, for example, of an injection moulded part.

Because the items are produced one at a time in a relatively labour-intensive manner, short runs in a wide range of self-coloured materials are feasible.

Due to the flexible nature of the tool, there is no requirement for draft angles on the model. However, an allowance for the natural shrinkage of the casting material, which can be considerable, must be added to the pattern before the tool is formed.

5.2.2   Disadvantages

Tool life is short (typically a few tens at best) and re-tooling requires the re-use of the relatively unstable SLA pattern.

The materials are relatively expensive and become scrap once the tool fails.
As with rigid tooling, changes are best handled with a fresh pattern and a full remake of the tool.

Size brings two problems – one being the physical limitation of the vacuum plant and ovens used – the other being dimensional stability and distortion of larger parts due to the difficulty of supporting and handling large, flexible tools in silicone rubber.

5.3 Immediate production tooling (IPT)

The form of tooling used by RIM-CAST in the case history quoted below is unlike either the *rigid* or *'soft'* forms described above and differs in the following respects:

- The tool is a hybrid in that the surface is of a compliant nature in polyurethane elastomer of a high wear resistance giving long life (from several hundreds to a thousand plus) without degradation.
- *The flexible surface also means that draft angles are not required on the model.*
- The main body of the tool, however, is a substantial, rigid, composite structure, stable in long-term production even for relatively large parts (up to 2 metres).
- For average-sized components a new tool is not only completed in a single working day but actually benefits from being used to make the first casting immediately it is finished.
- Loose pieces, side cores for undercut sections, and inserts can be incorporated in the same design philosophy that applies to injection moulding.
- Although a very small shrinkage allowance of 0.3 per cent is desirable, it is possible to reproduce models size-for-size on critical dimensions.
- Significant design changes can be introduced without a major re-tool since the solid nature of the tool body permits local areas to be removed and the surface re-cast to pick up the modification very economically.
- Because of the flexibility detailed above, immediate production tooling (IPT) yields production parts without the need for an intermediate prototype stage and can run for as long as the demand exists irrespective of quantity. The tooling example given below is for a product laid down nine years ago and this is still producing top-quality product on demand from the client. Over the life of the product to date more than ten detail changes have been introduced.

Fig. 1 Centre core of an IPT tool generating the body of the product illustrated

Fig. 2 Finished product originally rapid-tooled for production without 'soft' tool prototyping

6   SUMMARIZING TWO RIM-CASTING TECHNIQUES

Taking rigid tooling out of the frame because of the requirement of draft, which is seen as adding significant extra work at the design stage, it is not unreasonable to compare 'soft' silicone tooling with RIM-CAST immediate production tooling (IPT) on a 'horses for courses' basis.

Table 1*

Desired property	Silicone	RIM-CAST (IPT)	Comment
Rapid	Yes	Yes	One or two days
Low cost	Compared to rigid or metal tooling Yes	Compared to rigid or metal tooling Yes	Similar cost – silicone possibly cheaper for one tool only.
Thin sections	Excellent	*Practical limit on production 2 mm*	Possible but special care needed below 2 mm
Self colour	Yes	*Painting preferred*	Reason – variety of products on stream together
Fire retardant resins	Yes	Yes	Vac-casting resins expensive
High impact grades	Yes	Yes	Ditto
Production quantities	*Not usually unless very low*	Yes	IPT yields 5 items per day from single tooling
Retool cost	*Every 25 on an average product*	None for life of product	
Size	*Large becomes a problem*	No practical limit up to 2 m	
Design revisions	*Revised model and re-tool*	Incorporate with low or no cost	
Loose pieces, inserts, undercuts	*Possible but each added complexity can shorten tool life and impair quality*	All possible and accurately reproducible in long term production	
Item cost per moulding	*Prototype costing applies*	Commercial costing even for prototype quantities	

*Limitations given in italics.

7   FACTS

- Given a valid STL file, any of the free form shape-generating methods can produce a model in a small number of hours dependent on the item volume and size.
- A model maker can, equally, dress and finish the item in a few hours or less.
- The RIM-CAST (IPT) mould-making technique requires less than 24 hours to complete (usually less than 8 hours) creating a tool *and* producing the first-off casting within that time.
- The casting resins generally used on production cure in less than three minutes, which prompts the '*parts in minutes*' expression but is not the whole story. Size of component and shrinkage have a bearing on speed and accuracy.
- Realistically, CAD model to *first production batch* in two or three weeks is easy, in two or three days is possible, simple finished parts in a few hours from CAD model is feasible with careful planning.

Basically, the point being made is that technology, techniques, processes, and materials are in place to meet most required time-scale situations in rapid prototyping and production. However, in the vast majority of cases, it is the planning and attention to design quality that leads to success of any RP&M exercise.

8   GETTING THE DESIGN RIGHT

In numerous experiences with Rapid projects over the years, particularly when they start running late, one is tempted to issue a sticker for the design office which states:

> 'Please take the time to get it right. On a tight product launch date there may not
> *be* time to put it right when tooling is in place.'

This might seem tough on the designer or project manager but experience suggests that rapid technology should be viewed as:

> 'An intrinsic property of certain techniques which, if applied appropriately, results
> in benefits during product development.'

Rapid technology (or time compression, if preferred) should not be used as an excuse to leave major practical aspects of a project until the last minute. The product itself, however, at the end of the day is only as good as the design. Later in this article, a table illustrates how the early use of RP can refine a design and minimize problems – the thrust of which is to start using RP modelling, either SLAs or concept models from office modellers earlier in the programme, not just at the final stage of design.

9   SUMMARIZING THE ACTIVE SECTORS IN THE RP&M FIELD

Taking an overview of where the action is taking place in RP&M we have:

Sectors 1 to 10

Sector 1	Research into materials and techniques.
Sector 2	Material and equipment suppliers for RP generation purposes.
Sector 3	Bureaux offering RP services from model creation through to replication in various forms of tooling.
Sector 4	Design houses with RP and model creation capability.
Sector 5	Software houses serving the CAD/CAM/STL Industry
Sector 6	Large manufacturing companies with in-house RP facilities.
Sector 7	Rapid machining facilities with CAD/CAM and reverse engineering.
Sector 8	Rapid tooling facilities with on-going manufacturing capability.
Sector 9	New generation pattern/model makers with in-depth appreciation of RP techniques.
Sector 10	Finally, *the client*, whose markets in turn drive the industry.

(The position occupied by RIM-CAST incorporates activities in Sectors 1, 3, 8, 9, and, of course, Sector 10.)

Focussing for a moment on Sector 10, we find that whereas for a practitioner involved on a day-to-day basis with RP, all of the above aspects blended together form the somewhat eclectic way of life that is RP&M, in an increasing number of cases, although clients have an *awareness* of the availability of RP, they expect their suppliers as practitioners to be *expert* on the current developments because:

(a)   The client may only use RP for one new product every year or so.
(b)   He or she may not need to know every detail of the very latest processes so long as the budgets/timescales and technical objectives are all met.
(c)   The client may devolve responsibility to his design house/tool maker/moulder/RP&M service provider by simply defining the product conceptually or as a solid model in CAD and be prepared to simply pay for a result within a time-frame.

10   TABLE OF ENABLEMENT

In moving from CAD model to market place with the experience of over 200 completed projects, a pattern of designer thinking and procedure emerges which has been noted and summarized over the years, culminating in a simple check list.

Most popular stages of a new product launch appear in Table 2. The stages shown in columns 1 and 3 are client/designer driven and the options in the centre column are the RP&M options *enabling* the product to develop with the appropriate evaluation at each stage i.e. from top left to bottom right a.s.a.p.
It is possible that in 5 to 10 years time the centre column will disappear *for some products* designed with a new philosophy, where end-user products are produced directly by

stereolithography machines without the need for tooling, prototypes, or intermediate process steps. *In the meantime* the objective is to:

- Proceed from CAD to production in the most efficient way.
- If CAD data or 2D drawings are not available, to reverse engineer forms of tooling from models.
- If RIM-casting is the chosen process, to reach Point 2F (production) rapidly and economically.

Table 2

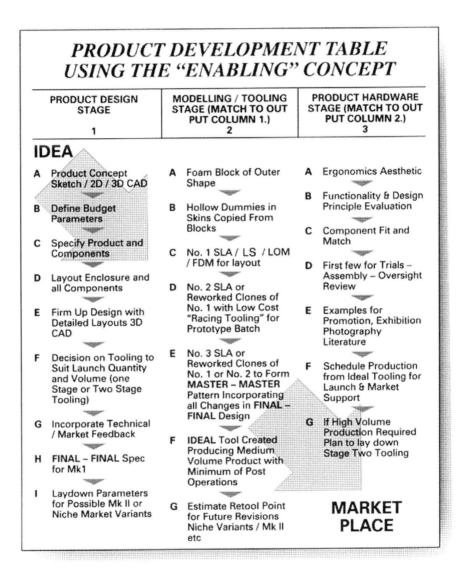

PRODUCT DEVELOPMENT TABLE USING THE "ENABLING" CONCEPT

PRODUCT DESIGN STAGE 1	MODELLING / TOOLING STAGE (MATCH TO OUT PUT COLUMN 1.) 2	PRODUCT HARDWARE STAGE (MATCH TO OUT PUT COLUMN 2.) 3
IDEA		
A Product Concept Sketch / 2D / 3D CAD	A Foam Block of Outer Shape	A Ergonomics Aesthetic
B Define Budget Parameters	B Hollow Dummies in Skins Copied From Blocks	B Functionality & Design Principle Evaluation
C Specify Product and Components	C No. 1 SLA / LS / LOM / FDM for layout	C Component Fit and Match
D Layout Enclosure and all Components	D No. 2 SLA or Reworked Clones of No. 1 with Low Cost "Racing Tooling" for Prototype Batch	D First few for Trials – Assembly – Oversight Review
E Firm Up Design with Detailed Layouts 3D CAD		E Examples for Promotion, Exhibition Photography Literature
F Decision on Tooling to Suit Launch Quantity and Volume (one Stage or Two Stage Tooling)	E No. 3 SLA or Reworked Clones of No. 1 or No. 2 to Form **MASTER – MASTER** Pattern Incorporating all Changes in **FINAL – FINAL** Design	F Schedule Production from Ideal Tooling for Launch & Market Support
G Incorporate Technical / Market Feedback	F IDEAL Tool Created Producing Medium Volume Product with Minimum of Post Operations	G If High Volume Production Required Plan to lay down Stage Two Tooling
H FINAL – FINAL Spec for Mk1		
I Laydown Parameters for Possible Mk II or Niche Market Variants	G Estimate Retool Point for Future Revisions Niche Variants / Mk II etc	**MARKET PLACE**

10.1 Using the table

Example 1 By the time 1F is reached 2D should have enabled 3A to D to be completed.
Example 2 Alarm bells should ring for instance if stage 1F apparently due to the product-to-market time scale, when exercise 2C has been left to the last minute and has just demonstrated that 3C is flagging up major problems (and, to make matters worse, the Marketing Director has only just seen the model, doesn't like the aesthetics, and wants to re-visit 3A!)

10.2 Applying the table to a case history

In the very specific example given next in this paper as a typical case history, the designer's confidence level in the chassis, designed and assembly-checked in Mechanical Desktop, was such that the plan followed became:

- 1A straight through to 1F
- 2E became the *first and only model* for tooling, was made by stereolithography from which 2F was created
- 3E, F and G followed in quick succession.

10.2.1 The product

The product illustrated in Fig. 3 is part of a DNA analysis system, the need for which was immediate and on-going, with production forecasts in the low hundreds.

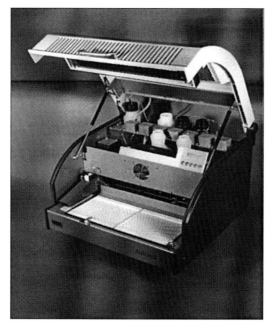

Fig. 3 The complete DNA analyser (to view this figure in colour, go to colour section)

The manufacturing philosophy, which worked well in the final analysis, was as follows:

- Metal formings were to be used for the base, structural sides, and the mounting for some of the heavier electrical components.
- *The main chassis and lid had to be very precise structural mouldings acting as anchor points for the majority of the functional components. Reagent bottles, peristaltic pumps, motors, and other components generated a multitude of critical interfaces.*
- Of prime importance was the precision with which the automated pipette arm could index to each selected sample-well to deliver or remove reagent.
- The alignment and support system for the large and extremely complex chassis comprised three metal rods CNC-located from wall to wall of the metal side cheeks. (Figure 3 shows the location of one of these in blue. The moulding would pick up its datum from these rods.
- The lid moulding also had multiple functions carrying displays and a heater element. (Figure 3 with the lid open shows the final effect.)
- Finally, moulded side cheeks enhanced the cosmetics and lid function. These carried large integrally moulded hinge pegs obviating the need for separate mechanical hinges.
- The completed production item to be built around the precisely moulded chassis.

Beyond the ability of any chosen process to meet the engineering precision, strength, and quality requirements, it was considered imperative that:

(a) Without repeated or further tooling investment it must be feasible *to proceed directly to medium volume production* in one single step, and

(b) Should design changes emerge in the transition from a *CAD-proven virtual model* to the *market ready article*, they could be taken on board even post-launch without major expense or delay.

10.2.2 Using STL files

Prior to tooling the chassis, the STL file (illustrated in Fig. 4) was of immense value in examining the geometry for undercuts and contra-drafts. Also a virtual tool was created using Materialise software and although the RIM-casting process is essentially a pattern replicating procedure, the virtual tool was instructive to mould design in studying:

- The vulnerability of the tool surfaces to potential damage in long-term use.
- Location of split lines for post dressing and aesthetics.
- Loose pieces and insert locations.
- Access to both halves of the tool for release agent application.

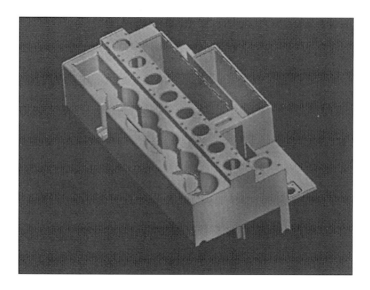

Fig. 4 The main chassis component in STL format (to view this figure in colour, go to colour section)

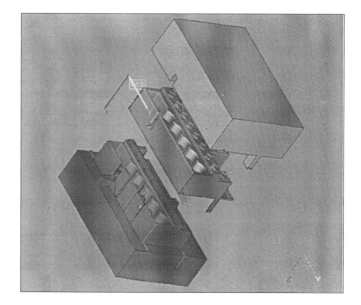

Fig. 5 Virtual tooling created in Magics RP (Materialise)
(to view this figure in colour, go to colour section)

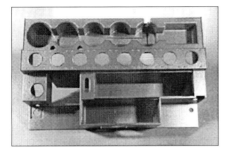 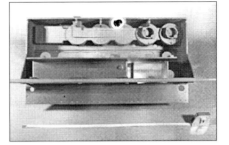

Fig. 6 The completed chassis casting----   **Fig. 7 CAD to production in one step**

10.2.3 Project status

The DNA Analyser has now been successfully launched and production is under way having avoided the multiple model/temporary tooling/non-production prototype stages which used to occur traditionally in product development using silicone tooling. Also, by using the RIM-CAST (IPT) process it was found unnecessary to go to high-cost metal tooling in order to achieve the accuracy, repeatability, and production capability that the product demanded.

The total product development time was 6 months and this was approximately one-third of the time usually associated with a product of this nature and complexity. The time-scale for the full set of moulded parts from data to production parts was under three weeks.

The combination of CAD, rapid prototyping (for the most complex model), immediate production tooling (IPT from RIM-CAST) and the design team's skills in project management enabled an ambitious launch deadline and on-going production needs to be met.

In terms of the Development Table the product is now into Stages 1I, 2G and 3G.

It is hoped that the information shared in this paper is of some value to new RP&M initiates.

Rapid tooling – cast resin and sprayed metal tooling

Joel Segal
School of Mechanical, Materials, Manufacturing Engineering and Management, University of
Nottingham, Nottingham, UK

ABSTRACT

Two broad classifications of rapid tooling techniques are direct and indirect. Direct
approaches use a rapid prototyping based process to make tooling inserts directly, whereas
indirect methods use the RP process to generate a pattern from which the tooling inserts are
made. This paper describes two such indirect rapid tooling processes – cast resin and sprayed
metal tooling. Case studies for each of the processes are presented and assessed in terms of
their impact on the product development business through cost and time savings.

BIOGRAPHY

Joel Segal received a bachelor's degree in mechanical engineering and a master's degree in
advanced manufacturing technology from the University of Manchester Institute of Science
and Technology. He then worked for two years as a research assistant in the Centre for Rapid
Prototyping and Manufacturing at the University of Nottingham developing sprayed metal
tooling for injection moulding. After a brief spell at PERA, he moved to the Rapid
Prototyping and Tooling team at Rover as a project engineer in the research team with
responsibility for sprayed metal tooling development and implementation. After nearly three
years with Rover, he is now working on a PhD at the University of Nottingham looking at the
effects of RP technologies on the properties of prototype parts for injection moulding.

1   INTRODUCTION

Although great strides have been made in rapid prototyping systems and materials towards
producing stronger, more accurate, and better surface finish prototypes, there is a great
demand for tooling to produce prototypes in the production material and using the production
process. Such prototypes are often referred to as *technical prototypes*. If these technical

prototypes are to be produced in short time scales then low-cost and reliable rapid tooling technologies are required.

Two broad classifications of rapid tooling techniques are direct and indirect. Direct approaches use a rapid prototyping based process to make tooling inserts directly, whereas indirect methods use the RP process to generate a pattern from which the tooling inserts are made. This paper describes two such indirect rapid tooling processes – cast resin and sprayed metal tooling. The processes are assessed in terms of their impact on the product development business through cost and time savings.

2   INDIRECT TOOLING – CAST RESIN AND SPRAYED METAL

Cast resin and sprayed metal tooling have been in existence for over 20 years but their reliance on the availability of accurate patterns somewhat limited their application. Since rapid prototyping (RP) systems became available, offering fast production of accurate parts that can be used as patterns for tooling, there has been renewed interest in these already well-established technologies. Both techniques, when coupled with RP technologies, can produce prototype tooling for processes such as injection moulding, compression moulding, and blow moulding in very short time frames and at very low cost. The low cost can mainly be attributed to the low cost of the materials used to manufacture the tooling. Typically both processes require skilled or semi-skilled labour and hence the labour cost may be relatively higher than other rapid tooling techniques.

Cast resin and sprayed metal tooling can both handle complex split lines well and undercuts can be accommodated by including machined metal inserts. On a like-for-like basis, sprayed metal tools can usually produce a higher quantity of parts than corresponding cast resin tools.

2.1   Cast resin tooling
Often referred to as aluminium-filled epoxy tooling or composite tooling, cast resin tooling is a process for manufacturing tooling using an epoxy resin as the tool material.

2.1.1   *Tool-making process (depicted in Fig. 1)*
The RP model is prepared on to a model board to define the negative of the first half of the tool. This is referred to as the pattern. An aluminium- or ceramic-filled epoxy is then cast on to the pattern directly in a supporting frame. The resin is allowed to cure (ranges from a few hours to over 36 hours). Further curing at elevated temperatures (up to 150 °C) can be used if increased strength is required. The process is repeated for the second half of the mould. When both halves have been cast and cured, the tool can then be split and prepared for moulding. Ejectors and runners are typically milled into the tool to ensure total functionality, and the tool support frames are integrated into a mould base for assembly on to the moulding machine.

Fig. 1 Resin casting process
[source: Process Chains for Rapid Technical Prototypes (1)]

2.1.2 *Advantages*

Cost – this is typically less than 40 per cent of the conventional prototype tooling cost.

Time – typical lead times lie between two and four weeks.

Surface reproduction – fine detail such as graining or print can easily be reproduced.

(a)

(b)

**Fig. 2 (a) Air ducting component; (b) Welding the modified section into a production
component (source: Bertrandt)**

2.1.3 *Limitations*
When moulding components with high aspect ratios, machined inserts are often required.

Flash can occur more often than with conventional tooling, leading to increased effort in trimming the mouldings.

Packing pressures are typically lower than those used in conventional moulds, which can lead to problems with sink marks.

Cycle times are longer than with conventional tooling and range from a minimum of 30 seconds for very thin parts to 1–4 minutes for products having more than 3 mm wall thickness.

Accuracy is dependent on a multi-step process, including the accuracy limitations of RP systems (0.05 mm at best). For this reason, when predicting the overall accuracy additional shrinkage and deformations have to be taken into account.

2.1.4 *Developments*
Developments of chemically bonded ceramics, special grades of resins, and specific surface coatings will help to further enhance the capabilities cast resin tooling. Also, a great deal of research is being carried out into how the properties of the parts produced from such tools varies from conventional tooling, primarily this research is targeted at the relatively low thermal conductivity of the cast resin tooling materials.

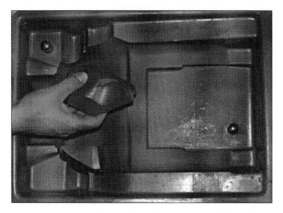

Fig. 3 (a) Cast resin tool cavity and loose resin insert (source: Bertrandt)

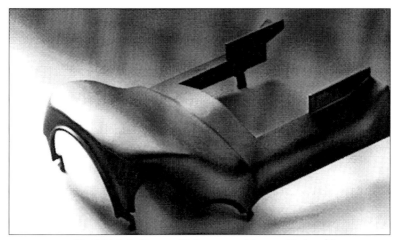

Fig. 3 (b) Injection-moulded component (source: Bertrandt)

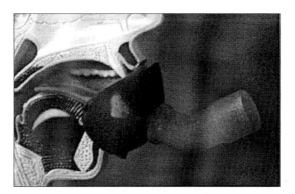

Fig. 4 (a) SLA and vacuum cast assembly of respirator mask for its design evaluation (to view this figure in colour, go to colour section)

Fig.4 (b) Cast resin cavity of mask

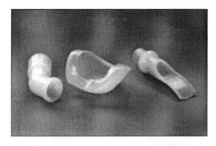

Fig. 4 (c) Injection-moulded prototype components of respirator mask

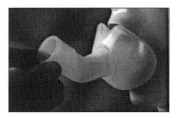

Fig. 4 (d) Testing an assembled prototype of the respirator mask

[Parts (a) to (d) of Fig. 4, source: Bertrandt]

2.2   Sprayed metal tooling

Metal spraying is a phrase used to describe, in general terms, a three-part process for depositing metal on to a substrate. An ENERGY SOURCE (e.g. electric arc, combustion flame, plasma flame) is used to melt a METAL (in wire, powder, or ingot form) and then a GAS (e.g. compressed air, nitrogen) atomizes the molten metal and propels it on to a substrate. In the case of sprayed metal tooling, the substrate is known as the pattern.

2.2.1   *Tool-making process*

The most popular process for making tooling is arc-spraying of zinc alloy, see Fig. 5. It was the first of all the currently available sprayed metal tooling processes to be made commercially available in the 1970s. Although the concept of sprayed metal tooling is not new, the interest in the technology has been re-kindled by the advent of rapid prototyping because of the availability of fast and relatively low-cost patterns for mould making.

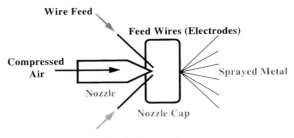

Fig. 5 Arc-spraying

Sprayed metal tools are produced by spraying atomized droplets of molten metal on to a pattern to form a metal shell (the face of the tool, Fig. 6). The pattern can be made from many different materials including most RP materials. The metal shell is sprayed to a thickness of approximately 1.5 mm and then backed up in a bolster using an aluminium-filled resin system.

Fig. 6 Arc-sprayed metal tooling

As the zinc alloy melts at 422 °C, it is classed as a relatively low melting point alloy. The small particles produced at the tip of the spray gun cool very quickly and this makes the material suitable for spraying on to stereolithography patterns, but it can also be sprayed on to other pattern materials such as wax, plaster, and even wood without problems.

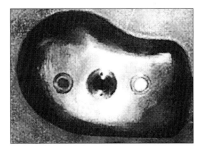

Fig. 7 Close-up of sprayed metal tool showing ejectors and sprue (source: PERA)

2.2.2 *Advantages*
Cost – this is typically less than 50 per cent of the conventional tooling cost.
Time – typical lead times for sprayed metal tooling are 35 per cent of the conventional lead time.
Surface reproduction – fine detail such as graining or print can easily be reproduced.

2.2.3 *Limitations*
All metal spraying processes suffer from limitations in geometry that can be sprayed because there are line-of-sight issues. For example, small deep features usually require inserts. The main limitation of arc-sprayed zinc alloy tooling in particular is tool life [Segal and Cobb (2)]. The numbers of components that may be produced depends on the manufacturing process, the moulding material, and the part size and geometry.

2.2.4   *Developments*

Because sprayed metal tooling is something of a 'black art' there is still some experimental and practical work to be done to investigate failure modes and optimize shell thickness, backing system, and basic tool design to develop a repeatable, clearly defined rapid tooling method for the production of prototype components.

Significant advances have been made in the last few years by a number of organizations in developing steel spraying for both prototype tooling and production tooling. The main advantage of steel over zinc alloy is that it is a much more durable material and it can withstand higher temperatures in the mould. Consequently, it is suited to more demanding processes, e.g. die casting.

Work has been done at PERA in Melton Mowbray, England, to produce arc-sprayed steel tooling using a single arc-spray gun and RP patterns (see Fig. 8). The steel is sprayed manually using a specially developed nozzle, which is currently being patented. The residual stresses in the steel shell can be controlled and the metallic backing system enables operation of tools up to 500 °C. The Danish Technological Institute in Taastrup, Denmark, use a process based around simultaneous shot peening for stress relief of the steel shell (see Fig. 9).

Fig. 8 Steel-sprayed injection moulding tool (source: PERA)

Here a five-axis robot is used to manipulate the substrate, the arc-spray gun and the shot peener through pre-programmed paths. The application of high velocity steel shot is used to create a neutral stress situation. Acceptable patterns include metal, soluble ceramic, SLA and low melting point alloy. Inert gases are used to lower the oxide content of the shell where required. The Sprayform™ process now being developed by Ford Motor Company is a patented process which produces arc-sprayed steel production tooling using the natural volumetric expansion of certain phase changes in steels to counteract the normal tensile stresses induced in the shell.

Fig. 9 Stress-free steel spraying (source: DTI, Denmark)

CONCLUSIONS

This paper has described only two indirect methods of rapid tooling. Many others are available, both commercially and in the advanced stages of development. It is fair to say that most rapid tooling technologies commercially available are still under development, i.e. they can be used successfully today, but extensive development work is being carried out to further enhance and improve them.

Obviously these techniques force some degree of compromise, either with respect to quantity of mouldings or appearance. However, provided a good understanding of their real capabilities is achieved and a competent subcontractor is found, they can deliver excellent results.

When considering which technology to choose, the best advice is to review published case studies because most of the technologies vary in the results that can be achieved. Factors include the part geometry, the part material, and the quantities required. Usually, a single rapid tooling technology cannot satisfy all prototype tooling needs in a company. Consequently, it is advisable to consider a range of techniques in the same way that a toolbox contains several different tools for tackling the job in hand.

ACKNOWLEDGEMENTS

I would like to thank Bertrandt AG, Böblingen, Germany for the cast resin photographs and PERA, Melton Mowbray, UK for the sprayed metal tooling photographs.

REFERENCES

1   Process Chains for Rapid Technical Prototypes, Brite Euram Project, Contract Number: BRPR-CT96-0155, Project Number: BE-2051. (http://ikppc43.verfahrenstechnik.uni-stuttgart.de/raptec/index.htm)

2   Segal, J. I. and Cobb, R. C. (1995) Optimising arc-sprayed metal tooling for injection moulding. Proceedings of First National Conference on *Rapid Prototyping and Tooling Research*, Buckinghamshire College, UK, November 1995.

Direct rapid prototyping (RP) tooling for injection moulds

T J Illston[†] and S D Roberts[‡]
†Land Rover, UK and ‡Warwick Manufacturing Group (WMG), University of Warwick, UK

1   INTRODUCTION

As market forces continue to push companies toward increasingly rapid product development, the demand for technical prototypes earlier in the design cycle has grown. This, along with a rise in the demand for niche products, has stimulated the development of rapid tooling technologies, where the cost and time required for the manufacture of technical prototype parts are significantly reduced.

The continued development of rapid prototyping techniques and materials has enabled these processes to be used to directly build tool inserts for injection moulding. Rapid tooling (RT) enables the manufacture of components during the prototyping stage in the final production material using the production technique. Direct stereolithography (SLA) and selective laser sintering (SLS) tools have both been successfully used for tool inserts for injection moulding. Direct stereolithography tooling is based upon the direct AIM™ process: AIM™ stands for ACES (accurate clear epoxy structures) injection moulding. The introduction of the copper polyamide (PA) material, a composite of copper and polymer, has also led to the use of direct SLS tooling for prototype injection moulding. The only previous SLS material suitable for injection moulding was RapidSteel and this required considerable post processing to create a metal tool for low-volume production.

There are many factors to consider in deciding the most appropriate route for prototype tooling; these include cost, lead-time, the number of parts required, the final material for the parts, and the part geometry. In order to maximize the benefits in terms of time and cost reduction for this type of tooling, the tool design should be simplified as much as possible to reduce the downstream tool-making. This includes using a direct sprue injection on to the part or designing a runner and gate system into the tool and removing the ejector system and using manual ejection.

This article will illustrate with actual applications the continued development of direct SL and SLS injection mould tooling over the last three years at Land Rover. A number of tools of varying complexity have been produced and they demonstrate the potential applications for the technology, and highlight some of the key considerations for a successful direct injection mould tool.

2   GENERIC TOOLING PROCESS

A generic process for designing and manufacturing the RP inserts is applied to both SLA and SLS tools, as described in the case studies. The inserts were designed to fit into an existing bolster system and all the features required, such as injection runners and gates, ejector pin holes, and location dowel holes, are included in the design. The inserts are generally built as a shell of between 2.5 and 5 mm thick, in order to reduce the build time and cost. The SLA shells were built in the tooling build style with 0.05 mm layer thickness to minimize the stair-stepping, the CuPA was build with a 0.075 mm layer thickness for the same reason. The shelled inserts were backed up with a room-temperature-curing, aluminium-filled epoxy resin to prevent the curing exotherm from distorting the inserts. The aluminium filler improves the compressive strength and thermal conductivity of the epoxy backing and also helps to reduce any curing exotherm. The backed up inserts are hand benched to match the split face of the two halves together and assembled into the injection moulding bolster system and the injection sprue and ejector pins are added. The ejection of the parts is greatly improved by removing the stair stepping on the vertical surfaces, but this finishing was kept to a minimum. The setting of the injection moulding parameters is critical to the durability of the tool, slow injection speeds and pressures are used and the tool temperature is monitored to avoid a build up of heat in the tool. The tool temperature for the SLA inserts was kept to below 45 °C and below 65 °C for CuPA, by having extended cooling times and cooling the surface of the tool with compressed air between mouldings.

3   CASE STUDIES

3.1   Case study 1: Trim patch
This component was an ideal candidate for using direct RP tooling, with a simple geometry and a requirement for only 10 parts. A direct AIM™ tool was built in Ciba SL 5195 resin and a simple ejector system was included in the tool design. The required ten components were moulded successfully in talc-filled polypropylene with no tool deformation and the project was completed in five days.

Fig. 1 The trim patch tool and component

3.2   Case Study 2: Airbag clips

The development of a new airbag needed functional prototypes to test the airbag deployment. A direct AIM™ tool was chosen for moulding the components as the simplicity of the tool design required only one removable metal insert to form an undercut clip feature. A series of three separate tools was produced to manufacture iterations of the part design with over 100 parts in talc-filled polypropylene being produced. Some tool wear was noted around the parting line and increasing flash formed on the parts as the tool became heated and deformed. Failure of the SLA inserts occurred on one tool around the loose metal insert, but an effective repair was undertaken.

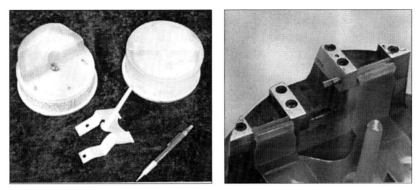

Fig. 2 Airbag clip tool with part and (right) close-up of repair to the damaged tool

3.3   Case Study 3: Steering wheel counter balance

Direct prototype tooling has also been applied to wax injection for the production of investment cast parts. A direct AIM™ tool was used to produce wax patterns for a brass steering wheel counter balance. The tool design was completed in Materialise Magics software to include the injector and locator for the wax injection. The wax was injected using the standard production methods, and checks on the initial samples highlighted a problem with filling of the tool, this required additional venting to be added. There were also some dimensional inaccuracies; these were determined to be due to using the same cycle times as an aluminium tool, with the parts being removed from the mould while they were soft. Using longer cycle times, to allow the wax to cool completely in the tool, solved the problem and 50 wax patterns were successfully produced.

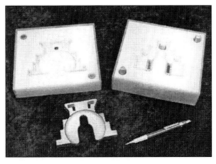

Fig. 3 Direct AIM™ wax injection tool and investment cast steering wheel counterbalance

3.4   Case Study 4: Glass guide

A prototype glass guide was required in the correct production material to check the packaging and operation of an electric window system. A copper PA mould tool was chosen as the most suitable route, as the Nylon 6,6 material and requirement for 100 parts was considered to be beyond the capabilities of a direct AIM™ tool. The tool design was generated for a standard bolster set for direct injection and a simple four ejector pin system was incorporated. The inserts were shelled to 2.5 mm thickness and produced on a Sinterstation 2500 system. The tool was used to produce a total of 117 mouldings in Nylon 6,6. Some gradual degradation of the tool occurred due to the high injection temperature of the polymer of 285°C. This resulted in localized melting of the sharp edges of the CuPA insert near the point of injection with radii forming on the part as shown in Fig. 4.

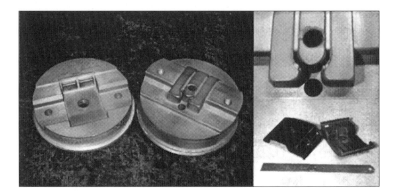

Fig. 4 The glass guide tool and part and (top right) close-up of the degradation of the tool

3.5   Case Study 5: Timing belt cover

A timing belt cover for a new diesel engine was required in the correct production material, talc-filled polypropylene, to allow the accurate measurement of the drive-by noise levels. Only one prototype was needed for testing and the relatively simplistic two-part tool geometry suggested that direct AIM™ tool was suitable. The tool design was simplified by having central sprue injection and tabs were added to the part to allow manual ejection from the tool. The 3 mm thick shelled tools were made using Ciba SL 5195 epoxy resin and built on a SLA5000. A simple bolstering system was used, consisting of aluminium side plates bolted together through the tool to support the clamping tonnage of the moulding machine and contain the lateral forces during moulding.

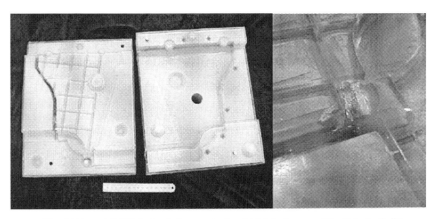

Fig. 5 Direct AIM™ timing belt cover tool and (right) close-up of the failure of ribs

A total of five mouldings were successfully produced within ten days from the tool from talc-filled polypropylene. Manual ejection using the tab features attached to the part worked but they increased the degradation of the tool by breaking off the weaker sections around the ribs as the part was removed. The use of manual ejection led to a long cycle time, exceeding five minutes, but this was not considered to be a problem when such a small volume of parts was required.

A subsequent requirement for 20 timing belt covers with a design modification resulted in a new tool being manufactured in CuPA. The CuPA tool was designed differently from the direct AIM™ tool and incorporated six ejector pins to ensure the part removal from the tool and the aluminium bolster was not used with the CuPA insert taking the full clamping pressure. A total of 30 parts was produced in the required talc-filled polypropylene to meet the functional testing with no tool degradation. To further test the durability of the CuPA inserts they were injected with a PC/ABS blend at 280 °C. A further 20 mouldings were produced but the tool was degrading with every shot due to softening of the insert during injection.

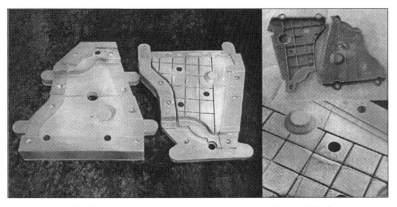

**Fig. 6 The CuPA timing belt cover and parts, and (bottom right) close-up
of the insert degradation**

4   CONCLUSIONS

Direct RP tooling, if applied correctly, can have significant time and cost savings. The case studies demonstrate that direct RP tooling can be successfully used to produce limited numbers of parts for functional testing. The benefit can maximised using tool design packages (for example 'Magics' from Materialise and 'Powershape' from Delcam) to generate tool data rapidly and by simplifying the tool design to reduce the downstream toolmaking thus allowing parts to be produced within as little as a week. Direct RP tooling is particularly suited to the manufacture of smaller tools, as the size increases alternative prototype tooling routes become more economic and effective. The success of these methods is heavily dependent on the part geometry, for example thin stand-up features may need to be produced using metal inserts. The direct RP inserts are relatively fragile compared to cast resin or metal tooling and special care is needed during injection moulding.

Each of the direct RP tooling routes has its own merits and limitations and these should be taken into account when choosing which route to apply. One of the key factors to their success is integration, ideally all the experience in RP, toolmaking, and injection moulding should be under one roof. The application of direct RP tooling should continue to develop in the future with improved RP machine accuracy and the introduction of new SLA resins and SLS powder dedicated to tooling applications.

Rapid manufacturing of dimensionally accurate mould inserts and metal components by Direct Metal Laser Sintering (DMLS)

Jouni Hänninen, Jan-Erik Lind, Tatu Syvänen, Juha Kotila, and **Olli Nyrhilä**
Rapid Product Innovations, Rusko, Finland

1   INTRODUCTION

One of the main deficiencies in today's rapid tooling and manufacturing (RT&M) techniques is the capability of producing only near net-shape parts. Accuracy has thus been one of the biggest problems for most processes and therefore it should be the primary development objective for all technologies. The importance of accuracy in tooling applications is even more critical than in other rapid prototyping applications. The overall performance evaluation of an RT&M system should, on the other hand, focus on the following critical factors:

- speed;
- accuracy;
- surface quality; and
- mechanical properties.

Speed has to be considered as duration from the start of the whole project to a finished mould or moulded parts. This way the cost efficiency of an RT&M technique can be compared with conventional methods provided of course that the geometry can be fabricated with conventional methods at all.

Dimensional accuracy can be considered as the most important criterion. If an RT&M technique is used rationally the fabricated part should be net-shape. Cost efficiency is best achieved when the surfaces need no additional machining after the fabrication. This goes hand-in-hand with surface quality: If the surfaces require extensive finishing, the dimensional accuracy is lost. Therefore, only minor finishing or post-processing is acceptable, if any.

The fourth factor influencing the performance is the mechanical properties of the fabricated structure. Conventionally the structure is built to be as strong as possible, which is not the optimal way: The mechanical requirements of the mould depend on the number of parts to be fabricated as well as on the material to be moulded or cast. Therefore, in order to save time and energy, the material properties of the mould should be tailored to withstand only as much strain as necessary. Additionally, the strongest material should be deposited only where it is needed, based on known stresses inside the mould.

2   DIRECT METAL LASER SINTERING (DMLS)

Direct Metal Laser Sintering (DMLS) is a laser-based and layered rapid tooling and manufacturing process developed by Rapid Product Innovations (formerly Electrolux Rapid Development, Finland) and EOS GmbH, Germany. The basic principle of the DMLS is to fabricate net-shape metal parts in one single process. Tool inserts or components are built directly and fully automatically from metal powder without any polymer binders, which ensures high speed and accuracy as well as practically no shrinkage; typical accuracy of the sintered parts is +/– 0.05 mm. During the process, a thin layer of metal powder is spread over a previously sintered layer and the cross-sectional area of the product is processed with laser scanning. Combining process control with the non-shrinking powder mixture has one remarkable outcome: sintered parts are net-shape. Surfaces require minor post-processing after the sintering and with special precautions dimensional accuracy is retained even with post-processing. The properties of the DMLS materials can be seen in Table 1.

Table 1 Commercially available materials for the DMLS process and selected material properties

	DirectMetal™100-V3	DirectMetal™50-V2	DirectSteel™50-V1*
Layer thickness	100µm	50µm	50µm
Main constituent	Bronze	Bronze	Steel
Ultimate tensile strength	≤ 200 N/mm^2	≤ 200 N/mm^2	≤ 500 N/mm^2
Brinell hardness	90–120 HB	90–120 HB	150–220 HB
Min porosity	20 %	20 %	5 %
Max operating temperature	600°C	400°C	850°C

(*DirectMetal and DirectSteel are trademarks of EOS GmbH)

3   APPLICATIONS OF THE DMLS

3.1   Injection moulding

Injection moulding is the most common application area of DMLS. Approximately 200 prototyping projects have been carried out successfully at Rapid Product Innovations and today DMLS is used as a standard and reliable tooling method for functional prototyping and short run series production. A noteworthy additional advantage of DMLS is the possibility of using internal cooling channels, which can be designed to mould inserts. Designers can place the cooling channels in an optimal way and curved internal shapes are also possible. This can be important for products in which the cooling channels should follow the geometry of the product as closely as possible for attaining optimal injection moulding conditions. The technique eliminates the limitations of conventional machining especially in moulds containing deep slots and recesses. Series size of injection-moulded prototype parts is typically from 100 to 5000 pieces. However, the material properties of laser sintered tool inserts enable the production of even a hundred thousand plastic parts.

A typical lead time for plastic prototype parts is 1–3 weeks. Normally customers send a 3D CAD model of the product after which the tool is designed. Tool insert geometries are designed using CATIA solid modelling and consequently STL files are generated. After that the STL files are sliced and transferred to the DMLS machine as SLI files. Sintering usually takes 1–3 days depending on the size of inserts. Finally the mould is assembled and transferred to an injection-moulding machine. Figure 1 illustrates a typical project flow of injection moulding with the DMLS inserts.

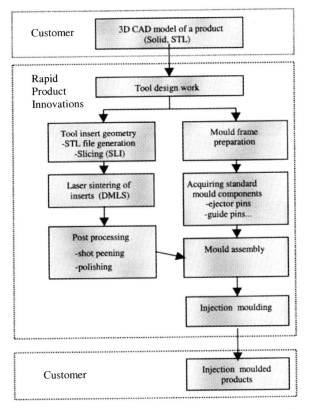

Fig. 1 Typical project flow of injection moulding with laser-sintered inserts

Case 'Twin Cover', a plastic prototype part for Volvo Car Corporation (VCC)

Volvo Car Corporation and its subcontractors became interested in testing the DMLS process in their development project after the first polyamide prototypes were fabricated using their own EOSINT P350 laser-sintering machine. After these first prototypes, more accurate parts were needed in the correct material, which was 10 per cent glass-filled polyamide and therefore the DMLS test tool was used. The project included three parts, one casing and two clips. Rapid Product Innovations carried out the project and delivered 200 parts to VCC. The lead-time of the project at Rapid Product Innovations was 2 weeks and the sintering time 40 hours. VCC compared the costs and lead-times of DMLS versus milling in this project and the results of the comparison have been published in the article 'Volvo's choice' in *Prototyping Technology International*. The results are summarized in Table 2 and Fig. 2 presents the component "Twin cover".

Table 2 Cost and time comparisons of the tooling techniques

Milling versus DMLS		
Case: Twin cover	*Cost*	*Time*
Production tool steel	EUR 75 236	14 weeks
Test tool aluminium	EUR 52 677	8 weeks
Test tool DMLS	EUR 20 433	2 weeks
Saved	**EUR 32 244**	**6 weeks**

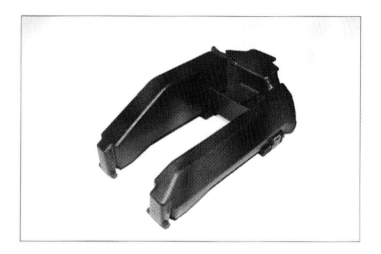

Fig. 2 'Twin cover', 10 per cent glass-filled polyamide

3.2   Pressure die-casting with laser-sintered inserts

Pressure die-casting is an application area where DMLS can give significant advantages prior to serial tooling. Processing conditions in casting are severe for the mould material and in particular higher operation temperatures limit the use of most RP materials. Studies and tests with magnesium and aluminium have proven that the DirectSteel™ material withstands even the extreme conditions of pressure die-casting. Improved material properties (tensile strength 500 N/mm², hardness 220 HB) and increased temperature resistance of the new steel powder enable the use of laser-sintered tool inserts in die-casting. Minimal post-processing is needed due to dimensional accuracy and excellent surface quality.

Case 'cutting arm', a chain saw component

A chain saw component 'cutting arm' was die-cast using laser-sintered mould inserts fabricated from the new steel-based powder (heat resistance up to 850 °C). The sintering time was approximately 60 hours/insert and 720 parts were cast. The inserts and cast parts are presented in Fig. 3.

Fig. 3 Die-casting inserts for the 'cutting arm' and cast parts

3.3   Direct steel component manufacturing: functional components even in 1 day

There are many cases in product development and the evaluation phase where certain functional features of a new product must be tested with prototype metal parts before the final decisions can be made. When only few functional components are needed there are not many options for manufacturing prototype parts quickly and cost effectively. Milling and investment casting are often used, but both techniques have some limitations and may not be time and cost effective. The new approach of using DMLS for fabricating steel components directly, has several advantages:

- complex parts can be produced directly from 3D CAD models;
- very short lead times possible (1–3 days);
- good material properties;
- good accuracy.

The latest developments of the DMLS process and steel powders for the process make it possible to manufacture complex metal components very fast. Geometries difficult or even impossible to produce by conventional methods, e.g. complicated internal channels, can now be produced in just a few hours and still maintain excellent material properties and high accuracy. The components are used for functional testing, for example in automotive and electronics industry as well as in household appliances industry. The applications are numerous; in the automotive industry laser-sintered steel components have been used, for example for testing gearboxes and engines.

The following pictures present components fabricated for functional testing. Dimensional tolerances and material properties have met the requirements of specifications and the components have been tested in real conditions. The components were sintered from the steel powder and the material properties were comparable to the final products. Delivery time of the gearbox component in Fig. 4 was only one day. The engine belt pulleys and the separator component presented in Fig. 5 were fabricated in two days.

Fig. 4 Gearbox component

Fig. 5   Engine belt pulleys and a separator component

4   SUMMARY

The progress of rapid prototyping and tooling techniques has been very intensive during recent years. In the beginning of the 1990s only light curable polymers and the stereolithography process were available. Parts were very fragile and applicable only to visual inspection. Today several polymers and metals including high melting temperature metals are available. The usability of the parts has improved and now even metallic durable steel tools and high-quality components can be generated directly from 3D CAD models.

Direct Metal Laser Sintering (DMLS) is a laser-based and layered rapid tooling and manufacturing technique originally developed for fabricating net-shape mould inserts for injection moulding. Development of advanced steel powders has widened the application area to pressure die-casting and direct component manufacturing. Complex tool inserts and functional metal components can now be fabricated rapidly in one single process. Improved mechanical properties and accuracy of the laser-sintered parts open totally new possibilities for utilizing RT&M effectively in product development projects.

Rapid laminated tooling

A L Threlfall
BAE SYSTEMS, UK

SUMMARY

Tooling manufacture is a significant element of aircraft manufacture in terms of both cost and lead time and a need has been identified by the aerospace industry to develop rapid, low-cost tooling. This paper provides an introduction to the laminated tooling development carried out within the LASTFORM programme. It also includes a case study detailing the manufacture of a laminated production tool for composite forming.

The work carried out to date demonstrates that there are substantial benefits to be gained by the aerospace industry from laminated tooling.

1   INTRODUCTION

Tooling manufacture is a significant element of aircraft manufacture in terms of both cost and lead time. Tooling currently accounts for around 40 per cent of the manufacturing development cost of a new aircraft design and, as the size of future aircraft batches is predicted to fall, the cost of production tooling will become increasingly significant. The lead times associated with very complex tools may be anything up to 12 months.

Tooling manufacture has therefore been targeted as an area for further investigation and development by BAE SYSTEMS and this resulted in the company leading the LASTFORM (Large Scale Tooling For Rapid Manufacture) programme. The objective of this programme was to develop rapid and cost-effective methods of producing tooling using novel layering techniques.

2   BACKGROUND

The concept of laminated tooling is very similar to laminated object manufacturing (LOM), the rapid prototyping process developed and commercialized by Helysis [1]. However, laminated tooling usually uses relatively thick metal sheets joined together rather than paper sheets bonded with thermoplastic. Interestingly, rapid laminating methods pre-date rapid prototyping by several years. The first known work on laminated tooling was published by Nakagawa in 1980 [2]. Since then many other researchers have entered the field of laminated manufacturing [3–6].

3 AEROSPACE MANUFACTURING PROCESSES

3.1 Range of processes

Laminated tooling development has been considered for a wide range of manufacturing processes used by the aerospace industry, from low-temperature injection moulding to medium-temperature composite forming and high-temperature superplastic forming. Work has also been focused on the large-scale tools required in the aerospace industry's drive towards unitisation and therefore larger and more complex parts. This paper is specifically focused on the development of tooling for composite forming.

3.2 Composite forming

Over the last 20 years there has been a significant shift within the aerospace industry towards lightweight materials such as carbon fibre reinforced composites. In addition to a high strength-to-weight ratio, the manufacturing methods for composite components allow several individual aluminium parts to be replaced by one single composite part. Although part integration provides major benefits in terms of a reduction in weight and assembly operations, it results in larger, more complex parts, which then present a major challenge in terms of the cost and time to manufacture tooling.

3.3 Current tooling processes

In this paper tooling for two commonly used composite forming techniques, autoclave and resin transfer moulding (RTM), will be described. However, many of the lessons learned will be equally applicable to other manufacturing processes within and beyond the aerospace sector.

In the autoclave process, composite material in the form of a prepreg tape (resin impregnated carbon fibre) is manually laid on to the surface of a single-sided tool. The composite is then enclosed in an airtight layer and placed under vacuum. The tool is then placed within an autoclave where it is heated to up to 200 °C and high pressure (6 bar) is applied to consolidate and cure the material. After the curing process has been completed the autoclave is allowed to cool to below 60 °C before removing the tool and stripping the part. To reduce the thermal mass of the tool and thus increase the speed of production, the tool is normally constructed as a thin shell between 8 and 20 mm thick. Typical manufacturing methods include electroforming nickel or fabricating from steel plates followed by machining. Both of these techniques are relatively slow and expensive.

Resin transfer moulding (RTM) is an important new process within the aerospace sector, which is capable of producing net shape parts with more precise dimensional control. In the RTM process a composite preform is placed into a matched die and then thermosetting resin (for example epoxy resin) is injected into the tool at relatively low pressure (2 bar). The resin infiltrates through the composite material and cures to produce the final part. To increase the speed of curing the tool is usually heated to between 150 and 180 °C. Traditionally tools have been machined from solid aluminium billets but this can be slow and expensive, particularly as the size and complexity of the tool increases.

4   LAMINATED TOOLING DEVELOPMENT

The main issues associated with the production of laminated tooling from metal sheets were identified as choice of laminate material, cutting techniques, build sequence, joining processes, and finishing techniques.

4.1   Joining techniques
The early stages of the programme were dedicated to evaluating a range of joining methods. The service conditions of the tools, the technical and economic feasibility, and the results of mechanical tests and cost modelling led to a down selection to adhesives, brazing or diffusion bonding, depending on the end user application. The experimental trials indicated that both RTM and autoclave tools could be successfully produced using either high-temperature adhesive bonding or a low-temperature brazing route.

4.2   Laminate cutting methods
A wide range of methods of cutting the laminations was considered, including water jet cutting, laser cutting, and high-speed routing. Although each process has advantages and disadvantages, laser cutting was eventually selected as a quick and simple option for cutting out the thin metal laminates.

4.3   Build strategy
Build strategy encompasses common factors such as *build sequence* and issues, which depend upon the particular tool geometry and application, such as *build orientation*.

Build sequence In the commercial laminated object manufacturing (LOM) process, each layer of material is stacked and bonded *prior* to cutting. This approach has the advantage that accuracy is ensured without the need for precise alignment of each layer. Unfortunately, though this method is very effective for paper, it requires blind laser cutting, which is difficult with metals and also presents a problem for subsequent removal of the waste material. The approach eventually selected in the LASTFORM Programme was *cut*, then *stack*, and then *bond*. Firstly, the profile is cut in to the lamination, including references points for subsequent alignment. The waste material is removed, the laminations are cleaned before any bonding agent is applied, and then the laminations are stacked to form the finished tool.

Build orientation The orientation of the laminations with respect to the tooling parting line is largely dependent upon the geometry and final application of the tool. For shallow tools it is usually more effective to laminate in the horizontal plane, as this requires a lower number of laminations. However, for complex tool geometry, horizontal laminating produces floating 'islands' of material, which must be located in some way (see Fig. 1). For deep tools, vertical laminations are favoured, due to the lower number of laminations that are needed. Moreover, tools constructed from vertical laminations do not normally generate 'islands' and thus this orientation is often favoured for complex tools.

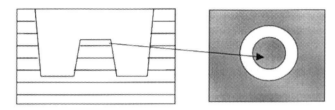

Fig. 1 Horizontal build orientation may cause 'islands'

4.4 Finishing methods

A number of alternative methods of reducing the stepped surface were explored during the programme including angled cutting, laser cladding, and laser edge welding [7–11]. However, the primary route ultimately used for the manufacture of tooling was simply to leave a machining allowance on the tool and then CNC machine the surface to produce the finished tool. This approach is a simple and effective method of producing the desired surface finish and accuracy, without the need for precise alignment of laminations during the bonding operation.

4. 5 Small-scale tool evaluation

The only reliable method of determining the performance of a new tooling route is to actually manufacture tools and test them. Small 'top hat' tools (see Fig. 2) were manufactured and evaluated under representative service conditions by the industrial partners. The results were very encouraging and it was decided to use the laminating process to manufacture larger tools – these are described in the following section.

Fig. 2 Small 'top hat' tools

5   CASE STUDIES

Having tested the performance of the alternative bonding mediums using small tools, the next stage of the programme was to manufacture representative tools. In addition to demonstrating the technical capabilities of the laminated tooling route, this part of the programme also enabled the cost and time of the laminating and conventional tooling routes to be compared.

In the case of the composite forming tools these were constructed from mild steel sheet, which was subsequently brazed together.

5.1 Autoclave tool

A hinge faring tool from Bombardier was selected for the autoclave tools trials. The existing CAD data for the tool were modified to incorporate a 2 mm machining allowance and the tool was 'hollowed out' to give a 20 mm thick shell. The revised tool design was then sliced into 3 mm thick layers, corresponding to a thickness of the mild steel sheet. The CAD model was sliced in several different directions and each layer was carefully checked (for islands and narrow features) to determine the most appropriate build orientation. Based on this evaluation a vertical build orientation was employed (see Fig. 3). To ensure alignment of the 91 separate laminations three steel bars, which pass through each lamination and are secured in 10 mm thick steel plates, were incorporated in the tool.

Fig. 3 Hinge faring tool before brazing

The completed laminations were prepared for brazing, stacked to form the tool, and a dead weight was used to apply sufficient pre-load to ensure that there were no gaps between the laminations during the brazing process. The braze conditions are shown in Table 1 below.

Table 1 Braze conditions – cowl tool

Braze material	Copper (99%)
Atmosphere	Hydrogen
Temperature (^oC)	1110–1160
Soak time (minutes)	30

After brazing the tool was machined by the industrial partner (see Fig. 4). The 'as machined finish' on the tool was excellent, with the laminations only being detectable on close visual examination (see Fig. 5).

Fig. 4 Hinge faring tool after brazing   **Fig. 5 Hinge faring tool after machining**

A comparison of the traditional and laminating route for the manufacture of tooling is shown in Table 2 below.

Table 2 Cost comparison for hinge faring tool

Tooling Route	Cost	Lead time
Conventional	£13 500 *	6 weeks
Laminated	£4 000 **	3 weeks
Saving	70%	50%

*Toolmaker quote ** Inc 50% profit

The cost savings for this tool are very significant. The conventional tooling route is that to machine the tool from a solid billet. This requires 80 per cent of the metal to be machined away to produce a shell tool with a low thermal mass. This is a slow and wasteful process. Firstly the inside of the tool is machined, followed by a stress relieving operation and then the outside of tool must be rough machined and then stress relieved, before the finished machining operation. For the laminated tool only final machining of the outside is required, without the need for any stress relieving operation.

5.1   RTM tool

An RTM tool (Coffin tool) was produced for Airbus, for use within a development programme. The CAD data for the tool were modified to take advantage of the laminating process. In addition to the 2 mm machining allowance, a complex network of heating passages was incorporated in the tool (see Fig. 6). In traditional tools the passageways would be restricted to straight drilled holes but the laminating process allows more complex heating circuits to be manufactured. For this tool the selection of the lamination direction was relatively simple. To reduce the number of laminations required and to withstand the pressure exerted on the tool it was decided to employ horizontal laminations. The tool was constructed from 60 mild steel laminations 3 mm thick, which were laser cut to shape. To ensure accurate alignment it was decided to incorporate two holes in opposite corners of each lamination, through which steel dowels were passed (see Fig. 7). The tool was brazed using a similar material and processing conditions to the autoclave tool (see Fig. 8). After completion the cavity was machined to size but the outside of the tool was left as laser cut to reduce the cost and time to complete the tool (see Fig. 9). One concern was the potential for leakage of the heating oil from the passageways, however, the tool was successfully tested and used to produce demonstration parts.

 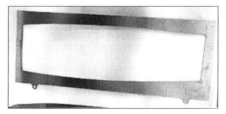

Fig. 6 Lamination showing complex heating circuit   **Fig. 7 Lamination showing dowel holes**

Fig. 8 Tool after brazing (section shows complex heating path)

Fig. 9 Inside of tool after machining

A comparison of the traditional and laminating route for the manufacture of tooling is shown in Table 3 below.

Table 3 Airbus – RTM 'coffin' tool (400 x 920 x180 mm high)

Tooling route	Cost	Lead time
Conventional	£11 350	8 weeks
Laminated	£4 730	4 weeks
Saving	58%	50%

The conventional tooling route for this tool is to either machine from a solid billet or to machine a near net shape casting. The majority of the savings for this tool result from the complex cooling passageways. These would normally be drilled into the tool, a slow and costly process, whereas for the laminated tool the passages can be laser cut during tool construction.

6   CONCLUSIONS

Laminated tooling is a viable alternative to conventional tool manufacturing routes. Significant cost and lead time savings can be generated but the degree of cost benefit has been shown to be configuration dependent, with some tool designs more suited to the laminated production process than others. The laminated process particularly lends itself to larger tools that would conventionally involve intermediate castings. Large cost benefits are also gained from the ability to cut and optimize integral heating/cooling channels in the metal sheets as conventional tool manufacture usually involves several drilling and filling operations.

When used effectively, laminated tooling offers the aerospace industry an opportunity to realize significant cost and lead time savings.

ACKNOWLEDGEMENTS

EPSRC (IMI), Warwick Manufacturing Group, Leeds University, Liverpool University, Airbus, Delcam plc, Quantum Laser Engineering Ltd, Rolls-Royce plc, Rover plc, and Bombardier.

REFERENCES

1   **Feygin, M.** (1988) *Apparatus and Method for Forming an Integral Object from Laminations*, US patent 4752353.

2   **Nakawa, T.** and **Suzuki, K** (1980) A low cost blanking tool with bainite steel sheet laminated. *Proceedings of 21st International M.T.D.R. Conference.*

3   **Dickens, P. M.** (1999) Principles of design for laminated tooling. *Int. J. Production Res.*, **35** (5), 1349–1357.

4   **Walczyk, D.** and **Hardt, D.** (1996) Recent developments in profiled-edge lamination dies for sheet metal forming. *Proceedings of the Seventh Solid Freeform Fabrication Symposium*, Austin, Texas, USA.

5   **Schrieber, M. P.** and **Clyens, S.** (1993) Blanking tools manufactured by laminating laser cut steel sheets. *2nd European Conference on Rapid Prototyping and Manufacture*, Nottingham, UK.

6   **Nakagawa, T.** and **Kunieda, M.** (1984) Manufacturing of laminated deep drawing dies by laser beam cutting. *Proceedings of 1st I.C.T.P.*, Tokyo, Japan.

7   **Erasenthiran, P., O'Neill, W.** and **Steen, W. M.** (1997) An investigation of normal and slant cutting using CW and pulsed laser for laminated object manufacturing applications. *EUROPTO '97 Proceedings*, Munich, Germany.

8   **Erasenthiran, P., Jungeuthmayer, C., O'Neill, W.,** and **Steen, W. M.** (1997) An investigation of step shaping using Nd:Yag laser for parts produced by laminated object manufacturing. *LANE '97 Proceedings*, Erlangen, Germany, pp 541–554.

9   **Erasenthiran, P., Ball, R., O'Neill, W.,** and **Steen, W. M.** (1997) Laser step shaping for laminated object manufacturing. *ICALEO '97 Proceedings*, USA, Vol. 83, Part 2, pp E84–E93.

10 **Erasenthiran, P., Clucas, D., Steen, W. M.,** and **O'Neill, W.** (1998) Material removal rate and depth determination in laser edge-matching. *ICALEO '98*, Florida, USA.

11 **Ball, R., O'Neill, W.,** and **Steen, W.** (1998) Laser Surfacing Techniques for Laminated Tooling. *ICALEO '98*, Florida, USA.

Section 5

Alternative Applications

Part I Medical Applications

Overview of medical applications

David Wimpenny

Using principles of rapid prototyping, CT scan data of anatomical parts, such as the skull, can be translated into data and used to manufacture a physical, three-dimensional (3D) model. There is a wide range of rapid prototyping machinery available, but the most commonly employed technique for medical models is stereolithography. Stereolithography models are ideally suited to surgical applications, as they are transparent, enabling even fine internal features including soft tissue to be accurately reproduced. This technique allows surgeons for the first time to hold accurate 3D models rather than have to undertake the painstaking task of interpreting 2D data from X-rays or CT scans. In addition to facilitating improved pre-operative planning, the models can also play an important role in achieving informed patient consent and are increasingly being used to improve communication in medico-legal cases.

Rapid prototype models can be used for pre-operative planning of these complex operations, resin replicas of the bones of the face may be re-fractured, repositioned, and fixed with titanium plates to replicate normal anatomy and eradicate facial deformity.

More recently, CAD methods have been used to design customized implants. These implants can be manufactured by investment casting from rapid prototype models directly or wax patterns produced using rapid tooling injection. Alternatively, the implant can be produced directly on high-speed CNC milling machines and on rapid prototyping machines in the future. Rapid prototyping techniques can be used to manufacture jigs in order to allow precise tissue resection and subsequent implant location. Using this approach, far higher levels of accuracy can be achieved than with hand and eye alone, leading to improved implant strength and aesthetics.

In this Section, the application of rapid prototypes and customized implants, from capture of patient data right through to manufacture of models, implants, and even tooling, is described. Kris Wouters from Materialise explains how scan data of patients are processed to enable individual features, such as tumours, to be isolated, allowing RP models to be produced. In this paper, he also provides an invaluable overview of the different medical applications of RP models and the relative importance clinicians place upon them. In the next paper, consultant surgeon Ninian Peckitt describes the use of RP biomodels and customized titanium implants in oral and maxillofacial surgery. In the final paper of this Section, the results of a research programme, conducted by Stryker/Howmedica Osteonics and Galway Mayo Institute of Technology, investigating the use of rapid prototyping and tooling technology to reduce the cost and time to develop and introduce cast orthopaedic implants, are described.

Medical models, the ultimate representations of a patient-specific anatomy

Kris Wouters
Materialise Medical, Leuven, Belgium

ABSTRACT

In the preparation of complex surgery, rapid prototyping (RP) represents a possible breakthrough. RP techniques allow the construction of real three-dimensional (3D) objects starting from computer data. Medical RP models or shortly medical models are physical hard copies of a patient's specific anatomy visualized by 3D scanning techniques (CT, MRI). These medical models provide visual and tactile information for diagnostic, therapeutic, and didactic purposes.

The CT Modeller System (**1**), a direct interface between scanner data and RP, that was developed by Materialise in cooperation with the Katholieke Universiteit of Leuven (Belgium), makes it possible to model selective parts of the human anatomy in a user-friendly way and even on a common personal computer.

Medical models are a very promising application of human visualization and have been used in both European and American hospitals for several years with remarkable results.
This paper will focus on the software side, i.e. the CT Modeller System, and will explain the different fields of application of medical models together with the first validation results. By means of introduction, the first section will focus on the medical advantages of some RP techniques. Colour stereolithography will be discussed in more detail, since it has some very typical medical applications (see Fig. 1).

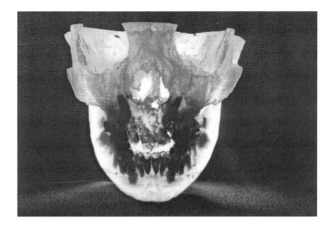

**Fig. 1 Skull model with teeth marked in colour; colour stereolithography technique
(to view this figure in colour, go to colour section)**

1   RP TECHNIQUES FOR THE PRODUCTION OF MEDICAL MODELS (2)

At the current time, RP has plenty of applications in industry, but its use to model parts of a
human body is rather new and still unknown to many people working in the medical sector. In
essence, all RP techniques can be used to produce medical models. However, some techniques
have important advantages over others.

Concept modellers of the 3D printing and 3D plotting type can be regarded as interesting
techniques for the production of medical models. These models are often quite voluminous,
and therefore the speed of concept modellers will press overall costs. The accuracy of a
concept modeller is still sufficient for most surgical applications.

FDM is less rapid compared with 3D printing and 3D plotting, but has some special
advantages. The technique is clean and easy to use, and this, combined with the fact that there
is a machine available with FDA approval and a material available that is 'medical grade',
makes FDM a suitable technique for placement in a hospital environment.

Stereolithography is less fast and definitely not suitable for placement in a hospital, but opens
the way to colour stereolithography. Colour stereolithography can be performed on any
stereolithography machine – it is only the resin used that is special. This resin called Stereocol
(a trademark of Zeneca Specialties, UK) is an FDA USP Class VI approved material that can
be selectively coloured, sterilized, and brought into contact with body fluids during the
surgical intervention. The colour is introduced by exceeding a certain dose of UV radiation,
higher than the dose used for initiating the polymerization reaction. Colour stereolithography
is still quite new, but already it is used by clinicians to highlight special structures selectively.

2   THE CT MODELLER SYSTEM

The CT Modeller System is a software system for interfacing from a medical scanner (mostly a CT scanner) to Rapid Prototyping or to CAD systems. The software is FDA approved and used by numerous service bureaux all over the world to produce medical models for local hospitals. Those service bureaux have an RP machine at their site and act as 3D printing offices. The CT Modeller software package that was developed by Materialise in co-operation with the Katholieke Universiteit Leuven is split in two main parts. This makes it possible for medical image interpretation to be performed by medical doctors, and technical processing of the models can be done by technicians. As indicated in Section 1, some machines are also suitable for placement in a hospital, where, driven by the CT Modeller system, they can act as the real 3D division of a radiology department.

Patient data are imported in the software where a selection (segmentation) of the region to model is made. The relevant voxels are selected from the huge set of images. After the segmentation, the software is used to calculate the machine files necessary to build the medical model. By interpolating the image data, accuracy of the model is enhanced to sub-voxel level and staircasing effects are avoided. The time needed to build a model starting from the raw data coming from the scanner is often less then 24 hours. No expensive 3D workstations are needed and the accuracy can be quite high, depending on the distance between the slices.

3   VALIDATION

In the PHIDIAS project (**3**) (a European project on medical modelling), a validation questionnaire was developed by a radiological team from the Katholieke Universiteit of Leuven. Based on this questionnaire, stereolithography models and the cheaper, faster, and more easily obtained alternative of static pseudo 3D images (shaded images) were made. Different types of case were selected in which a model can be used. Sixty-eight models were produced for 48 patients. The most important results can be summarized by two graphs, see Figs 2 and 3.

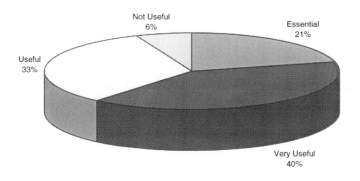

Fig. 2 Overall assessment of usefulness of SL models based on a validation of 48 patients by 25 surgeons

The first one reflects the overall assessment of the usefulness of the SL models reported on the Phidias study. In comparison to pseudo 3D images, the medical model led to different decisions in many of the cases that were studied. A graphical interpretation is given in the figure below.

The Phidias network, a networking project sponsored by the EC and started 1 April 1998, can be regarded as the successor of the PHIDIAS project. In the Phidias network, the validation study will be extended in number (670 cases planned) and will be performed on a European scale (11 different European countries participating).

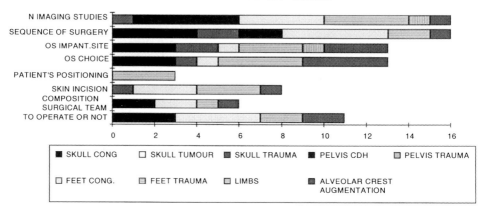

Fig. 3 **Figure indicating in which cases the use of models did lead to a different decision**

4   USE OF MEDICAL MODELS

Medical models are an excellent tool for the visualization of a patient's anatomy. Two-dimensional pictures or 3D shaded images can be very hard to understand, even for experienced medical doctors. A real size model of the patient's pathologic region gives tactile information and makes it very easy for the doctor to understand the physical problem. The model helps him to gain better insight into the complexity of the patient's problem. Real time rotations are evident and the communication between the doctors and patient is facilitated in a high degree.

Apart from the visualization and communication, the models are often used for the planning and even rehearsing of complex surgery.The rehearsal on a model can be performed using the same surgical tools as during the actual surgery, while having an overview that is impossible during the actual surgery. By rehearsing, the surgeon reduces greatly the risks of surgical surprises.

Colour stereolithography allows selectively colouring of regions inside a medical model, e.g. tumour tissue, dental roots, etc. Therefore, colour models are even better for visualizing and planning in certain cases. Also, in implant surgery, medical models have a wide scope of applications, where they can serve directly as a master for, or as a negative of, the implant.

5   CONCLUSION

The software system presented makes it possible to model any part of a patient's anatomy that is clearly visible on the scanner images. The so-produced medical models have the potential for saving time, saving expense, and improving the surgical results.

Future developments are concentrated on the production of surgical templates, soft tissue organs out of CT or MRI scans, and expanding the interfacing between all kinds of medical scanners and all kinds of rapid prototyping systems.

REFERENCES

1   **Swaelens, B.** and **Kruth, J. P.** (1993) Medical applications of rapid prototyping techniques. *Proceedings of the Fourth International Conference on RP*, Dayton, Ohio, 14–17 June 1993, pp. 107–120.

2   **Rapid Prototyping Association** (1997) RP Technology: a unique approach to the diagnosis and planning of medical procedures, pp. 15–17 (Rapid Prototyping Association of the Society of Manufacturing Engineers).

3   **Materialise NV, Siemens Medical, Katholieke Universiteit Leuven,** and **Zeneca Specialties** (1996) PHIDIAS project summary report.

Rapid prototyping in the biomedical industry

Stryker/Howmedica Osteonics, Limerick, Ireland

ABSTRACT

A buzzword that has been around for some time now is rapid prototyping, also known as desktop automated manufacture and free form fabrication. Since industry in general has become more aware of rapid prototyping and tooling (RP&T) there has been a growing demand for it to be used to decrease the time taken for product development and to increase the accuracy of engineering data. This in itself pays dividends by saving the manufacturer money, getting their products to markets earlier, eliminating costly mistakes, and gaining significant competitive advantage in the market place.

This paper aims to show how to adapt RP&T technology and techniques to facilitate the creation of patient specific implants, complex assemblies, regional products, clinical evaluations, and product launches.

1   INTRODUCTION

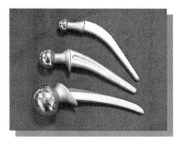

Fig. 1

Stryker Howmedica Osteonics based in Limerick, Ireland is a leading orthopaedic implant designer and manufacturer of implants for the replacement of diseased joints of the human body. Examples of hip implants are shown in Fig. 1 and knee implants in Fig. 2 (left). A sophisticated prescriptive service also operates on site providing custom-made implants for oncological applications as well as other specific patient needs such as joint revision surgery.

Fig. 2

Since its opening in 1971 the Limerick plant has produced in the region of 300 000 knee replacements for implantation around the world. All the cast products manufactured in the plant are of the patented Vitallium™ cobalt–chromium–molybdenum alloy and there is a precision casting facility which allows control of the entire manufacturing process, from casting to final packing. It is also the dedicated site for the production of Surgical Simplex® bone cement [see Fig. 2 (right)], the most widely used bone cement in the world for prosthetic implant fixation.

Computer aided design and rapid prototyping are highly utilized for form functionality testing and aesthetic qualities as well as mould manufacture and CNC programme generation. The 3D CAD systems that are currently being used are Unigraphics and Pro/Engineer. A Stratasys FDM machine is used for the majority of rapid prototyping needs and MIMICS software enables the conversion of MRI CAT Scan data into solid 3D CAD models for pre-operative walkthroughs and research and development (R&D). For quality checking, PC/DMIS CMM reverse engineering software is available.

2   FROM PRODUCT DEVELOPMENT TO FINISHED PRODUCT

2.1   Rapid prototyping in product design
Currently RP is changing traditional approaches to the design, production, and launch of a global range of products. Ideas for new projects and products can come from many sources but are usually originated from R&D, marketing or product development groups. Design of a product will normally be undertaken with the assistance of a collaborating surgeon or group of surgeons. The first step of any project usually involves the design engineer visiting the collaborating group to discuss the project requirements. The engineer then starts the initial design, literature searches, initial CAD modelling, prototyping, and initial FEA. The use of RP allows early visualization of a design for the surgeon, which also allows him to make any design changes that may be necessary.

Figure 3 gives a good indication of the life cycle of the design / manufacturing process from start to finish. From the initial 3D CAD model a rapid prototype is generated and this is then shown to surgeons and is used with mock bones and existing parts for fit and functionality testing. Once the product design has been approved a mould is designed and machined from aluminium or cast via a brass master pattern in bismuth-tin alloy.

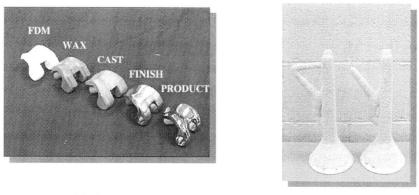

<table>
<tr><td>Fig. 3</td><td>Fig. 4</td></tr>
</table>

This mould is used for the injection of investment casting wax. The wax master is set up on a wax post, often with ash drains and vents and is then dipped in ceramic slurry to build up a shelled coat (see Fig. 4). This is sent to the autoclave to remove the wax pattern; once this has been done the shell is preheated for casting. After the part has been cast the ceramic shell is removed, the casting is leached in caustic sodium hydroxide to remove any ceramic cores that may be present. The parts are then cut off from the set-up posts. All operations that follow are a series of milling, turning, grinding, linishing, stoning, and belting operations that improve the overall finish on the final product. All products are ultrasonically cleaned and sent for passivation – which helps to make the components inert and helps them resist body reaction. The final stage in the process is the sterile packaging of all components.

3   RAPID PROTOTYPING APPLICATIONS

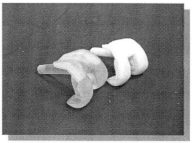

Fig. 5

A diversity of RP technologies and techniques has been tried and tested including sterolithography (see Fig. 5), fused deposition modelling (FDM) (see Fig. 5), fast casting urethanes, selective laser sintered rapid steel tooling, and casting parts from expendable RP patterns. In terms of RP, FDM plays a significant part in the activities of the company and the generation of business. The main reason for the purchase of such an expensive piece of RP equipment was to gain access to a physical part in a compressed time frame. Because of the level of accuracy of FDM parts and the innovations of many people – both inside and outside the company – the technology has expanded to other areas of application including rapid production tooling and direct pattern generation for product launches, and production of low quantity products. It was felt that there was a need for a clear visualization and communication of ideas between the design engineer and the surgeon in a three-dimensional format. For this reason RP is utilized wherever possible. Below are a number of small projects that have been undertaken to evaluate the use of different RP techniques.

3.1   Fused deposition modelling

Fig. 6

The FDM machine uses acrylo-nitrile butadiene styrene (ABS) to make prototypes, some of which are shown in Fig. 6. This machine serves its purpose in bridging the gap between conventional two-dimensional drafting and three-dimensional CAD models. This gives the design engineer and the surgeon an actual physical prototype of a design, which both can hold and review without the large expense involved in machining a component from metal. Intricate geometries, complete assemblies, snap, locking, and threaded features have been made possible in current designs. Cadaveric trials are used to evaluate parts of the design for actual implantation in the human body. In this case FDM models were used in conjunction with finished product to evaluate the final assembly. It can be said that FDM has considerably reduced development lead time on new products and constantly facilitates the early detection of possible design faults.

3.2   SLA rapid production tool

Fig. 7

The most recent and direct application of rapid production tooling was the investigation into the possibilities of using a SLA (ACES build style) solid model to mould wax patterns for investment casting. The device to be manufactured is a non-sterile, referred to as 'trial', prosthetic implant. The implant is used by the operating surgeon to check the 'fit' of the cavity that will be made to receive the prosthesis plus bone cement that will be used to fix the implant in place. The mould tool shown in Fig. 7 was created on Unigraphics 3D CAD by subtracting the original CAD of the medical implant from both of the mould blocks. The original 3D CAD model was scaled 2.3 per cent larger to accommodate wax and casting shrinkage. The SLA mould was shaped so that there was adequate strength and rigidity built in. Unnecessary material was removed to reduce the overall cost of the SLA part. Orientation of the sterolithography model proved an important factor. In an effort to eliminate the occurrence of stair stepping on both halves of the mould tool, both halves were generated standing in an upright position. It was observed that due to the thermal properties of the SLA resin the tool would heat excessively unless a very high cycle time was used. However, with induced cooling (a blast of air) the cycle was reduced but was taking between one and two minutes. The outcome of the project was that the appropriate numbers of trial parts were produced without the need for expensive dedicated metal moulds.

3.3   SLA vacuum casting masters

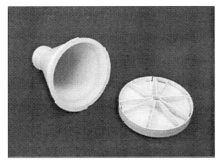

Fig. 8

The device in Fig. 8 is a sterolithography model of a disposable surgical cement mixer. This SLA master has been post processed and painted with exceptionally good results. The painted finish significantly reduced stair stepping and has improved the overall surface finish on the SLA model significantly. Extra clearance had been allowed on the original CAD model for the application of a painted surface, which was estimated to be typically 0.3 mm. The SLA model exhibited good dimensional accuracy, and this was reflected on special features such as the thread for the mixing bowl and a self-sealing lock design on the lid of the model. This SLA master has been used to make a silicone mould for the vacuum casting of marketing samples for customer preference trials. This resulted in the manufacture of 20 samples that were provided to the customer to functionally test the product. The customer was actually able to mix with some of these samples. The SLA master has also been used to check the final fit on packaging materials. When in full production these parts will be injection moulded from polypropylene.

3.4   Direct and indirect casting from RP

Currently an on-going RP project in house is investigation into the possibilities of casting small lots of custom medical implants *directly* and *indirectly* from RP&T. Small batch quantities, cost, and lead times are the main contributing factors for the purpose of this investigation. This is the basis of current research activity within the organization.

To produce parts *directly* means that a RP model will be used instead of the wax pattern in the investment casting process. To produce parts *indirectly* means that a mould, which has been generated from RP&T techniques, will be used to manufacture wax patterns that will in turn be used to investment cast the parts.

3.4.1 Direct route for casting

Figure 9 shows expendable selective laser sintered patterns. These patterns will be used in the place of the wax pattern in the investment casting process. Prior to the manufacture of the prototype, the original CAD file had been scaled by 2.3 per cent to accommodate the overall shrinkage in the wax and metal. Burnout procedures are to be evaluated with regard to time and the amount or residual ash left behind in the ceramic shell prior to casting. All of the prototypes involved in this test (Fig. 10) have undergone various finishing techniques in order to improve the surface finish prior to casting. Parts are to be evaluated for surface finish dimensional variation and will undergo x-ray inspection. The models being used in the casting trial are:

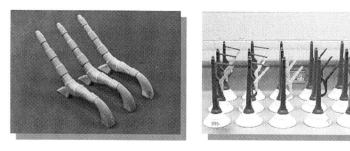

Fig. 9   Fig. 10

- selective laser sintering, Trueform (Fig. 9)
- multi jet modelled, Actua thermoplastic
- sterolithography, Quickcast
- fused deposition modelled, ABS
- layered object manufactured, paper
- ballistic particle modelled, Sanders thermoplastic.

3.4.2 Indirect route for casting

Producing investment casting patterns by RP can be successful if only a few parts are needed. As the number of prototypes increase, RP becomes increasingly expensive and more time consuming. 'Soft tooling' can be useful as many wax patterns can be produced relatively quickly and cheaply once the injection tool is made.

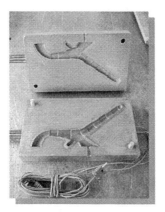

Fig. 11

Two approaches have been taken. An aluminium-filled epoxy resin tool (shown in Fig. 11) was manufactured using a SLS prototype as the master pattern. Shrinkage compensation factors were applied to the SLS model. An overall contraction allowance of 2.3 per cent was applied to the SLS master pattern to accommodate shrinkage in the wax and casting stages. This epoxy resin tool incorporates cooling lines and thermocouples inserted at specific points to record the effect of temperature when the wax is being injected. This tool has yielded successful waxes with little or no flash around the parting line and the cycle time is approximately 30 seconds per shot of wax with an air blast through the integrally moulded copper cooling lines.

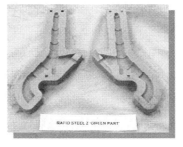

Fig. 12

The second approach shown in Fig. 12 is an SLS rapid steel tool in its 'green' or semi-sintered state. Here stainless steel powder coated with a plastic binder is fused together by a laser beam. The part is now in its 'green' or 'semi-sintered' state. This is then placed in a furnace, which debinds the steel powder and lightly sinters the particles. The resulting 'brown' part is put within a graphite crucible and is then infiltrated via capillary action with bronze, resulting in a fully dense steel–bronze composite tool. The original 3D CAD file is used as input into the SLS machine. The CAD file has a shrinkage allowance of 2.3 per cent made for both the wax and the Vitallium™ metal. The moulds being evaluated for the wax injection trial are:

- aluminium cast epoxy resin tool (Fig. 12 left)
- selective laser sintered rapid steel tool (Fig. 12 right)

- room temperature vulcanizing silicone tool
- sterolithography tool

4   CONCLUSION

The above examples show how the company is currently utilizing rapid prototyping technology to investigate alternative methods of 'soft' and 'hard' tooling to further research and development into new projects, improve the design process as well as the launch and production of standard and custom implants. This paper has given a general overview and insight into the various applications of rapid prototyping and tooling within Stryker Howmedica Osteonics. With the proper application of rapid prototyping and tooling technology, lead times and costs in the design and production of joint replacement implants will be reduced. As the use of these technologies increases within the medical industry, additional applications will be discovered and utilized.

BIOGRAPHY

Kieran Walsh holds a degree in manufacturing and, while working at Howmedica Osteonics, pursued his masters degree in rapid tooling.

Daniel Boyle is a lecturer and researcher at the Galway-Mayo Institute of Technology. He studied engineering at NUIG and holds two masters degrees. Before lecturing he worked for 10 years in industry both in Europe and the USA for IBM, the American Crane Corporation, and the Copeland Corporation.

Rapid prototypes and customized implants in maxillofacial reconstruction

Ninian S Peckitt, FRCS, FFD RCS, FDS RCS
Consultant Oral and Maxillofacial Surgeon and Director, ComputerGen Implants Limited, UK

1 INTRODUCTION

Traditional head and neck surgery involves complex surgical reconstruction techniques, which do not replicate the volume and contour of normal anatomy. *'Functional reconstruction'* involves replication of the normal volume and contour of both hard and soft tissues, to produce normal form and function of the face mouth and jaws (1–4). Using rapid prototypes and customized titanium implants it is possible to simplify the surgical procedure following tumour ablation, for example, to a single stage operation, which provides full reconstruction including dentition, without the use of composite flap cover (2–4). This method of functional reconstruction reduces the severity and duration of operations, which benefits the patient by lowering surgical trauma. Moreover, this approach requires less theatre time and personnel, reduced ITU utilization, and leads to earlier patient discharge, enabling potential cost saving of between £17 000 and £19 000 per case to be realized (3, 4).

2 FUNCTIONAL RECONSTRUCTION

Conventional reconstructive surgery involves long, complex procedures (5, 6), which often require the harvesting of bone, muscle and skin, and its blood supply, from a second surgical site (7–15). This blood supply is reconnected in the neck using microvascular free flap transfer techniques. The dentition is reconstructed separately using titanium implants in a multistage procedure. This approach, which may entail surgery lasting 12–18 hours and involve multiple surgical teams, places the patient under high levels of surgical trauma thus increasing the risk of perioperative mortality, morbidity, and poor/delayed rehabilitation. *'Functional reconstruction'* is impossible to achieve with living donor tissue, especially in those cases involving replication of complex osseous anatomy. However, functional reconstruction can be achieved using rapid prototype biomodels and customized implants.

2.1 Rapid prototyping
Rapid Prototyping has been used in conjunction with CT, MRI, and even ultrasound scan data to produce anatomical biomodels. The scan data of the patient anatomy is transmitted using either T-1 Internet Links and DICOM protocols, or Optical (WORM) Discs. These data require reformatting prior to model manufacture. This post processing may be carried out

using software, such as Mimics from Materialise, which permits the setting of thresholds related to the x-ray absorption properties of the skeleton and overlying tissues. The software outputs an STL file of the desired biomodel, which is then constructed on a stereolithography machine, for example (see Fig. 1). This technology permits the accurate replication of any organ, and is invaluable in the planning of complex surgical procedures. The development of stereolithography resins, which change colour after being irradiated for a second time by the laser, has enabled tumours to be mapped accurately and is invaluable in the planning of complex surgical cases.

The introduction of rapid prototyping engineering technology into clinical care has made a major contribution to the description of injuries, treatment planning, and the efficacy of treatment, in:

- facial trauma;
- facial deformity;
- reconstructive surgery;
- medicolegal practice;
- craniofacial surgery.

This technology has a wide application across many surgical specialities:

- *Orthopaedics* – trauma management, customized jigs, joint prostheses, spinal surgery.
- *Neurosurgery* – surgery of nerve root pain, tumour resection.
- *Maxillofacial surgery* – oral rehabilitation, implants, jaw resection, and reconstruction.
- *Craniofacial and skull base surgery* – trauma, tumour resection, and reconstruction.
- *Plastic surgery* – scar revision, wide applications in reconstructive surgery.
- *Otorhinolaryngology* – head and neck reconstruction, nasal reconstruction.
- *Vascular surgery* – customized vascular stents.

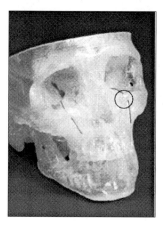

Fig. 1 SLA biomodel
Diagnosis and treatment of facial pain: facial fractures.
History of sensory nerve entrapment/numbness with pain (dysthesia). The nerve was freed from a
constricted bony canal using a precise cutting jig. Pain free at 3 months, with return of normal sensation

2.2 Customized implants

Having used a biomodel to plan the surgical procedure, the next stage of the process is the design and manufacture of a customized titanium implant. Again scan data from the patient are used to enable a titanium implant, which is a facsimile of the bone, to be designed. The implant is then manufactured, using CNC machining methods, to an individual prescription (see Fig. 2). The use of computer generated implants permits greater accuracy of replication of normal anatomical contour and jigs can also be manufactured which enable precise resection of tissues and location of the implant (see Fig. 3). The relationship between the implant and the overlying soft tissues can be assessed by digital photography and morphing programmes in a technique we have termed '*Photomorphanalysis*'. The implant is inserted and fixed to the skeleton using evidence based surgical techniques. The author has found that exposure of nasal and oral titanium surfaces without flap cover (1–4) is possible. This permits a single staged procedure, with preoperative manufacture of removable overdentures, which are secured to the implant by established precision attachment mechanisms.

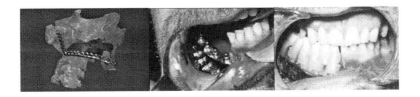

Fig. 2 (left) Customized implant mounted on biomodel, (mid) implant *in situ*,
(right) implant with overdenture at 57months

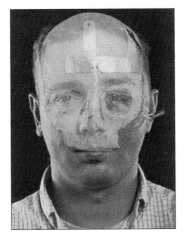

Fig. 3 Reconstructed shattered eye socket (orbit) using customized titanium
implant and jig for precise repositioning of cheek bone (shown using photomorphanalysis – the
superimposition of digital images)

3   APPLICATIONS OF FUNCTIONAL RECONSTRUCTION (16–21)

3.1   Medically compromised patients

Some patients have medical conditions, for example arterial disease, which normally would exclude them from major cancer surgery. Morbidity, related to the donor and recipient sites (15–19) is not uncommon. Partial or complete flap loss has been reported as high as 28.6 per cent (18) for free fibula transfer to 33 per cent (21) with respect to reconstruction of the extensive defects of the head and neck in a series of 648 patients. Recurrence of tumour may occur in the flap used for the reconstruction and this leads to salvage problems. Bone has the propensity to die after radiotherapy, and this results in the painful and distressing condition known as osteoradionecrosis.

Such medically compromised patients may be denied a surgical component to their treatment and ultimately succumb to their disease. The reduction of surgical trauma afforded by more simple surgical techniques and customized implants could influence the outcomes of morbidity and survival.

3.2   Resection of huge tumours

Farag and McGurk (15) have indicated that the risk of experiencing a complication increases two fold by each increasing magnitude of surgery, and that 'major procedures have four times the complication rate of minor operations'. Peri-operative mortality figures within the first 20 days (20) have been reported as high as 9 per cent, and are probably related to the metabolic response to surgical trauma. The use of customized implants and the reduced trauma of the reconstructive component of surgery is making the treatment of huge tumours possible with reduced risk for the patient as a function of reduced surgical trauma. The concept of customized implant reconstruction must be compatible with conventional methods of reconstructive surgery so that salvage is possible as a second stage procedure in the event of implant failure.

3.3   Facial trauma

Biomodels have been used to manufacture jigs for the accurate positioning of fractured parts of the facial skeleton. Customized implants have been used successfully in the reconstruction of the orbit (eye socket), cheekbone for augmentation and resection, and in the management of facial pain, where nerves can be accurately dissected from bony canals. The technology is particularly well suited to the management of facial asymmetry, which may be treated with implants manufactured from a mirror image of the contralateral side in very short, simple, and safe procedures.

3.4   Distraction osteogenesis

The use of biomodels will permit the design of customized devices to influence the growth of bones in the face using the principles of *distraction osteogenesis*. In this process bones can be sectioned and slowly jacked apart. New bone forms in the gap as the bone lengthens. Biomodels can be harnessed to influence the choice of vector with the production of an ideal contour and volume of bone, with minimal surgical trauma.

4   CASE STUDY: RECONSTRUCTION OF THE UPPER JAW (MAXILLA)

4.1   The traditional approach

Traditionally, reconstruction of the maxilla has involved mutilating procedures with compromized functional results. The Webber Ferguson surgical approach to the upper jaw involves dividing the upper lip in the midline, extending the incision lateral to the nose and then below the eye, so that the half of the face is opened like a book. One half of the maxilla can be resected using this approach. The maxilla is reconstructed with an upper denture on which is placed a bung (obturator) to fill in the huge defect left by the resection. Recently, Tideman (Hong Kong) has described a complex osseous reconstruction of the maxilla. The new maxilla is made from a titanium mesh tray, which is filled by bone particles taken from the hip and ground into a paste. The bone graft is covered by temporalis muscle, taken from the temple, which is harvested through an incision going over the top of the head (bicoronal flap). This muscle provides the environment for the ingrowth of blood vessels from the muscle into the graft, which survives, and revascularizes over a period of six weeks with minimal loss of bone. Titanium dental implants may be inserted into the bone graft for attachment of teeth or dentures. These implants fuse (osseointegrate) with bone and may be brought through the tissues to the external environment without the development of infection as described by Brånemark.

In the author's experience the Tideman procedure took 13 hours to complete and was associated with complications related to incomplete wound healing and partial loss of implanted bone and dental implants. The surgical trauma involved in this option is extensive, involving surgery at three sites, in the mouth, the hip, and the scalp.

4.2   Aesthetic maxillectomy

The problem of maxillary reconstruction has been greatly simplified with the use of a customized titanium maxilla. Tumour resection was planned on the biomodel and a customized maxilla was made from titanium alloy (1–4). The titanium maxilla is designed to be an anatomical facsimile of the resected parts (Fig. 4).

Flanges for fixation of the implant to the cheekbones were titanium welded on to the main superstructure. With an antero-posterior path of implant insertion, additional flanges into nasal bone undercuts provide resistance to displacement by the forces of gravity. Internal fixation to the facial skeleton was provided with 2 mm titanium screw fixation. Resection of the tumour was carried out through the mouth in a 2½ hour operation. The implant was secured to the cheekbones with minimal access, by passing a trocar through the cheek so that fixation screws could be fitted to the bones without opening up the face with a Webber Ferguson incision.

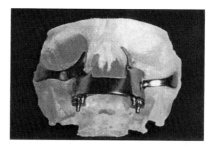

Fig. 4 Customized titanium maxilla

Four precision overdenture attachments were incorporated into the implant and an upper overdenture was made prior to surgery (Fig. 5). The complete manufacturing process was completed within 14 days from the date of initial consultation to the time scheduled for surgery, by a local orthopaedic engineering company. The overdenture was fitted at the time of surgery, and the patient returned to the ward. No intensive care facilities were required. Speech at 24 hours post operation and retention of the overdenture was excellent.

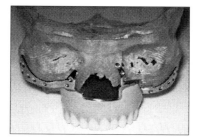

Fig. 5 Implant with overdenture (retained by precision attachments)

Anatomical implant loading would occur through the region of the base of skull, and also through the cheek flanges across to the top of the skull (calvarium). It was decided that the implant would not be loaded by an opposing denture for a period of six months following surgery, after which a conventional lower denture would be constructed, in an attempt to reduce early mechanical loading forces of the implant. No postoperative radiotherapy was required and the patient remains disease free with excellent function and aesthetics at five years (see Fig. 6). When the exposed maxilla was viewed with an intraoral mirror 12 months after the operation it was found to be clear of any debris, apart from soft tissue attachment with the oral mucosa around the periphery which forms a water tight seal. The nature of the attachment is unclear but thought to be related to glycoprotein secretion (see Fig. 7).

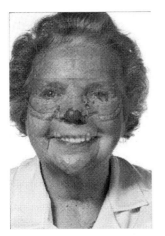

Fig. 6 Patient with implant *in situ* using photomorphanalysis

Fig. 7 Exposed titanium maxilla

4.3 The atrophic maxilla

If the entire maxilla can be reconstructed following resection it is possible to adapt this technique to reconstruct the wasted or atrophic maxilla, a common condition, which prevents adequate retention of the upper denture. These cases are extremely difficult to reconstruct and involve multistaged bone grafting procedures and the insertion of dental implants. This treatment is expensive and is associated with major complications that can result in the loss of the entire maxilla.

Rapid prototyping permits cost-effective surgery, with precision planning, and the manufacture of customized implants, which are cantilevered from the cheekbones, for attachment of the upper overdenture. This surgery is currently carried out in a 20 minute procedure under general anaesthetic

5   COST EFFECTIVENESS

In 1993 at the British Association of Oral and Maxillofacial Surgeons meeting in Cardiff, Lavery stated that the true cost of intra oral reconstruction utilizing free flap techniques was estimated at £25 000 per case. The cost of this case of 'functional reconstruction' utilizing stereolithographic model planning and a customized titanium implant manufacture in 1999 is £14 535 per case for purchasers. However, if one considers the two year mortality for advanced mouth cancer to be in the region of 70 per cent, the average cost of caring for a patient for a two year period is £85 000 for the free flap group assuming a survival rate of three out of ten patients. One unexpected finding that in a group of seven patients with large tumours treated with customized implants, there have been no deaths at four and a half years at a cost of £14 535 per case and further investigation is warranted on much larger groups of patients. With respect to the management of trauma cases savings are higher per unit case as a function of the consequences of reduced surgical trauma. Savings of £30 000 are possible with projections of reduced operating times, less dependency on critical care facilities, earlier discharge from hospital, and enhanced rehabilitation.

One case attending from the USA involved documented savings of $150 000 for the purchasers.

6   FUTURE DEVELOPMENTS (22)

This technology is being developed through the 'Medilink' Association with the aim of forming an European Institute of Biotechnology which will link engineers, physicists, metallurgists, biotechnologists, etc. with clinicians. The applications of this technology are enormous, and have special reference to oral and maxillofacial surgery, orthopaedics, neurosurgery, skull base and craniofacial surgery, spinal surgery, ENT and plastic surgery, vascular surgery, and veterinary surgery.

The introduction of this technology has far-reaching implications in the field of reconstructive surgery in general. Further development and investigation of this technology is advocated especially with respect to:

- implant biotechnology and design;
- the biological processes involved in this technology;
- the development of treatment planning protocols;
- associated effects of this technology on morbidity and perioperative mortality figures;
- projected estimations of reductions in hospitalization and rehabilitation times;
- quality of rehabilitation;
- projected estimations of overall cost savings.

ACKNOWLEDGEMENTS

The input from the multidisciplinary surgical and engineering teams based in Doncaster and South Yorkshire are gratefully acknowledged.

REFERENCES

1   **Peckitt, N. S.** (1997) Stereoscopic lithography and the manufacture of customised implants in facial reconstruction. *Brånemark Reunion*, Blackrock, Dublin, May.

2   **Peckitt, N. S.** (1997) Stereoscopic lithography and the manufacture of customised implants in facial reconstruction. Poster presentation: *Oral Diseases*, 5th International Congress on Oral Cancer, London.

3   **Peckitt, N. S.** (1998) Stereoscopic lithography: customised titanium implants in orofacial reconstruction. *European Association of Craniomaxillofacial Surgeons Meeting*, Helsinki.

4   **Peckitt, N. S.** (1999) Stereoscopic lithography: customized titanium implants in orofacial reconstruction. *Br. J. Oral and Maxillofacial Surgery*, **37**, 353–369.

5   **American Joint Committee on Cancer** (1981) *Manual For Staging Of Cancer*, Third Edition, Purposes and Principles of Staging (Eds O. H. Beahrs, D. E. Henson, R. V. P. Hutter, and M. H. Myers), Ch. 1, pp 3–10 (Lippincott Company, Philadelphia).

6   **Stell, P. M. and McCormick, M. S.** (1975) Cancer of the head and neck. Are we doing any better? *Lancet*, **11**, 1127.

7   **Raveh, J., Stich, H., Sutter, F.,** and **Greiner, R.** (1984) Use of titanium-coated hollow screw and reconstruction plate system in bridging of lower jaw defects. *J. Oral and Maxillofacial Surgery*, **42**, 281–294.

8   **Raveh, J., Roux, M.,** and **Sutter, F.** (1985) The lingual application of a reconstruction plate: A new method in bridging lower jaw defects. *J. Oral and Maxillofacial Surgery*, **43**, 735–739.

9   **Raveh, J., Sutter, F.,** and **Hellem, S.** (1987) Surgical procedures for reconstruction of the lower jaw using the titanium-coated hollow-screw reconstruction plate system: bridging of defects. *The Otolaryngologic Clinics of North America*, **20**, 535–558.

10   **Raveh, J., Vuillemin, T., Ladrach, K., Roux, M.,** and **Sutter, F.** (1987) Plate osteosynthesis of 367 mandibular fractures. The unrestricted indication for the intraoral approach. *J. Cranio-Maxillo-facial Surgery*, **15**, 244–253.

11   **Hellem, S.** and **Olofsson, J.** (1988) Titanium-coated hollow screw and reconstruction plate system (THORP) in mandibular reconstruction. *J. Cranio-Maxillo-Facial Surgery*, **16**, 173–183.

12   **Vuillemin, T., Raveh, J.,** and **Sutter, F.** (1988) Mandibular reconstruction with the titanium hollow screw reconstruction plate (THORP) system: Evaluation of 62 cases. *Plastic and Reconstructive Surgery*, **82**, 804–814.

13   **Vuillemin, T., Raveh, J.,** and **Sutter, F.** (1989) Mandibular reconstruction with the THORP condylar prosthesis after hemimandibulectomy. *J. Cranio-Maxillo-Facial Surgery*, **17**, 78–87.

14   **Raveh, J.** (1990) Lower jaw reconstruction with the THORP system for bridging lower jaw defects. *Head and Neck Cancer*, **2**, 344–349.

15   **Stoll, P., Bahr, W.,** and **Wachter, R.** (1990) Bridging of mandibular defects using A O-reconstruction plates. THORP – VERSUS 3-DBRP-System. Presentation: *10th Congress European Association of Cranio-Maxillo-Facial Surgeons*, Brussels.

16   **Karran, S.** (1982) Who needs nutritional support? In *Surgical Review* (Eds J. S. P. Lumley and J. L. Craven), **3**, pp 25–62 (Pitman).

17   **Farag, I.** and **McGurk, M.** (1997) A prospective study of complications encountered in the surgical treatment of oropharyngeal and maxillary cancer. *Oral Diseases* 5th International Congress on Oral Cancer. Programme and Abstracts, **3**, S8.

18   **Husseiny, M.** and **Fata, M.** (1997) Free fibula skin paddle. A way to improve its reliability? *Oral Diseases* 5th International Congress of Oral Cancer. Programme and Abstracts, **3**, S8.

19   **Webster, K.** and **Brown, A. M. S.** (1997) The versatility of the scapula flap in oral reconstruction. *Oral Diseases* 5th International Congress of Oral Cancer. Programme and Abstracts, **3**, S8.

20   **Herter, N.,** and **Novelli, J. L.** (1997) Lateral trapezius myo and osteomyocutaneous flap on head and neck reconstruction. *Oral Diseases* 5th International Congress of Oral Cancer. Programme and Abstracts, **3**, S62.

21   **Azizyan, R. I., Matyakin, E. G., Uvarov, A. A., Fedotenko, S. P., Kropotov, M. A.,** and **Podvyaznikov, S. O.** (1997) Plastic correction of extensive defects of head and neck. *Oral Diseases* 5th International Congress of Oral Cancer. Programme and Abstracts, **3**, S63.

22   **http:// www.maxfac.com**

Section 5

Alternative Applications

Part II Architecture and Art

Overview of architecture and art applications

Chris Ryall and Julia McDonald

For most engineers, designers, and now even surgeons, rapid prototyping is nothing more than a tool which, when used appropriately, can save time, money, and effort while improving the final result.

This Section looks at the impact of rapid prototyping on the arts and architecture. These are relatively new areas for rapid prototyping and it should be noted that there are plenty of other areas, which although not covered by this book are beginning to emerge. Two such areas are archaeology and palaeontology where scanned data have been processed to reveal three-dimensional forms of such diverse objects as the remains of a sunken ship and a fossil.

Rapid prototyping has a major potential for manufacture of customized products. Unfortunately, direct manufacture of end products by rapid prototyping in traditional engineering applications is slow to be adopted. Artists, however, are in a unique position to utilize the rapid prototyping processes to produce the final work of art directly. Keith Brown's contribution highlights this unique situation. Keith's preconceptions of rapid prototyping led him to view the modelling process as little more than a means to the production of a bronze sculpture. However, Keith comments that when seeing the pattern for the first time he was 'truly amazed by the breathtaking appearance of the object....it was not just a prototype'. The production of the model raised an awareness of the potential of the RP processes themselves as a media for art. This discovery raises a plethora of issues that would have otherwise not been given a second thought for any other application. Many bureaux and rapid prototyping engineers have developed a range of finishing techniques to eliminate the step effect on the surface of models. Within the art environment, the stepped or faceted surface may represent an integral feature of the final object. Artists might also wish to retain other features, such as the support structure, which they feel enhance the overall effect. It has only been through the introduction of RP into art, by activities such as the CALM Project, that RP engineers have become aware of these issues. The CALM project, which is described by Elizabeth Hodgson, has also helped to make artists more aware of the need to produce data in the right format. For the digital artist, emphasis has previously been solely on generation of the visual effect within the computer environment, not the technicalities of generating a sound CAD model. In

Section 1, Peter Dickin covers issues of CAD generation that are applicable to all users generating data for output to rapid prototyping processes.

An alternative approach, avoiding the need for an artist to use CAD directly, is the use of reverse engineering. Chris Lawrie describes several case studies illustrating the application of reverse engineering within the arts. Using these techniques, free-form organic shapes can be captured, scaled, and manipulated before manufacture.

Utilizing rapid prototyping in architectural modelling presents a whole range of unique issues and problems. Like engineers, architects require visualization models for communication to contractors and customers. However, functionality is rarely required. Charles Overy discusses some of the unique problems that occur due to such issues as the scale of architectural models to perhaps one hundredth of their actual size. The economics of producing architectural models by rapid prototyping is also seen as a restricting factor.

Traditional model-making techniques, combined with the practice of utilizing junior/training architects to manufacture models, mean that the cost (true or perceived) of using the existing methods seems to be much lower. Perhaps another important factor is that traditional architectural modelling techniques are learned and seen as an integral part of the training and development process.

In conclusion, for the full benefits of rapid prototyping to be gained in these new and exciting applications, it is vital for the RP industry to understand fully the requirements of artists and architects and that they in return appreciate the particular characteristics of the RP processes.

Sculpture for the new Millennium

Keith Brown
The Manchester Metropolitan University, Faculty of Art and Design, Department of Fine Arts, Manchester, UK

ABSTRACT

Recent developments in three-dimensional (3D) modelling applications and RP have facilitated completely new ways for sculptors to develop their practice. They are now becoming increasingly aware of the revolutionary possibilities being made available through RP. Many users can now be found in the Computers and Sculpture Forum in the USA, Ars Mathematica in France and FasT-uk (Fine Art Sculptors and Technology in the UK). Examples and illustrations are given, including a description of the process of designing an RP as a master model for the lost wax bronze casting process.

1 NEW TECHNIQUES, NEW FORMS, NEW SCULPTURE

Until recently it would have seemed like magic for the fine art sculptor to be able to conceive of and manifest their ideas using a microcomputer and 3D physical output devices. What many engineers are now beginning to take for granted has only very recently become a reality for the artist. Sculptors are only just becoming aware of the power RP offers as a viable technology for the production of their work.

Recent development of inexpensive powerful microcomputers and the proliferation of affordable, user-friendly, software packages for 3D modelling have brought this technology to the desktop and at last into the studio of the sculptor. With the release of every new version of software, developers are improving the ease of use by making the graphic interface and the behaviour of modelling tools ever more intuitive. For the most part, perhaps this is of more importance to sculptors than it is to product designers, architects, or engineers, who now have a tradition of using CAD as a means to design their product. For many sculptors CAD is a much less intuitive process than it is for the designer and may often seem somewhat cold and off-putting to the fine artist. The introduction of these more direct and user-friendly software applications has helped the sculptor to overcome this barrier. The new, very powerful, highly responsive, real-time, rendering engines allow for the direct manipulation of forms in 3D space. It is not so long ago that rendering was a separate element in the modelling process and very difficult to use. It was something like modelling in the dark. The process was akin to turning the lights on to see the rendered model, off to make adjustments, and on again to see the result. However, it is now possible, with most 3D packages, to be able to attribute textures to objects and to see them with shading, in real time, as you modify and work with them. Much of this WYSIWYG software development has been driven by the world of multimedia, for application in film and video special effects and computer games. It certainly has not been developed specifically for the fine art sculptor but has offered them a new medium with techniques to create 3D objects, and concepts that are leading to the emergence of completely new types of art.

1.1 Alternatives to modelling
In addition to modelling directly on the computer, there also exists the possibility to acquire 3D data via 3D input devices in the form of 3D digitizers and 3D laser scanners. In addition to this a vast amount of ready-made material is available on the World Wide Web in the form of 3D clip art. The latter lends itself readily available to a kind of computer junk art, assemblage, or bricolage, for which I have coined the term cyberlage as a suitably descriptive term. The size and scale of these objects are instantly adjustable and make it possible to bring together 3D entities from as far apart as the micro- and macrocosms in a way never before imagined. It is an easy matter to combine, say, a red blood cell with the latest 3D data model of the universe and fuse them together into a single object, should you have reason enough to do so. These works may then be output to a rapid processing device and thus become manifest as tangible objects. It has never before been possible to acquire readily available and accurate models of this kind. Although work being done in this area is still in its infancy, the possibility to acquire and use this data is at our fingertips. There are also various other input devices, many of which are still in R&D, that employ force feedback modelling tools from data gloves to stylus that respond to and mimic prescribed material properties, and which promise to make the modelling process even more directly intuitive for the fine artist. We already have a wide range of modelling tools including software-clay, metaballs, metacloth, NURBS, etc.

2   ANTICIPATING A SUDDEN AND MASSIVE DEMAND

It is perhaps not surprising given the relative newness, availability, and high cost of RP technology, that there are but a few sculptors to date that have ventured into the realms of RP. During the 1990s a number of sculptors' organizations were formed around the use of the computer and related technologies. Between them they have done a great deal to raise awareness of RP in the fine art community. During 1992 Alexandre Vitkine and Christian Lavigne combined forces to found Ars Mathematica in France. In the USA a group of digital sculptors formed a loose-knit community called the Computers and Sculpture Forum of North America, which was developed around the International Sculpture Centre in Washington D.C. during 1992 under the direction of Rob Fisher, Timothy Duffield, Bruce Beasley, and David Smalley. In the UK, and in part as a direct result of the Higher Education's Funding Councils CALM (Creating Art With Layer Manufacture) project, FasT-uk (Fine Art Sculptors and Technology in the UK) was founded by me December 1997.

All of these sculptors' organizations started out with just a handful of members. Growing in strength throughout the 1990s their collective membership must now be in the region of about 200–300. In the UK the CALM project (mentioned elsewhere in these pages by Dr Elizabeth

Hodgson) saw a doubling of the world number of sculptors using RP in the short space of just two years. Altogether, to date, there must be about 50 sculptors worldwide who have experimented with RP and produced work in this way.

As awareness of the possibilities being offered by RP spreads throughout the fine art sculpture fraternity, more and more sculptors are realizing the unique features of this technology. Intersculpt, an international biannual symposium of digital sculpture organized by Ars Mathematica since 1993, will be hosting the first international competition for rapid prototype sculpture. From an open international submission TeleSculpture works will be built, in real time, at the various galleries around the world during the duration of the exhibition.

2.1 New materials and techniques
For the sculptor as opposed to the engineer or product designer, RP might well be considered an inappropriate term. For many sculptors who have used the technology, the objects that result are not considered as prototypes, but as art objects in their own right. Since the Romanian sculptor Constantine Brancusi coined the term 'Truth to Material' in the early days of Modernism, the appropriateness of material and the part that this plays in sculpture has contributed a great deal both in the conception and production of their art form. It is the use of the microcomputer as a design tool, combined with the RP processes and materials available to make their creations manifest as 3D objects, that seems to offer the greatest advantage to the sculptor. This is over and above any consideration for the speed of production. It is the promise of a truly new way to conceive of and produce their art where this technology comes into its own as a challenging and exciting new medium.

Fine Art sculptors are free individuals when it comes to their art form and they do not need the constraints of a design brief. One of the general concerns of the artist is an ambition to contribute to the body of knowledge that we call art by bringing into being that which has, as yet, not existed and to derive from it or attribute to it some kind of meaning and understanding. This places them in a unique position as users of this technology in what is otherwise a world of engineering and science. Acknowledged, of course, is the enormous debt to science and engineering for inventing and developing these processes in the first place and without which the artist would not be in the position to consider using this technology at all. However, it is the very high cost of producing work in this area that perhaps most stunts their imagination and contribution. Of the few sculptures created in this way, in most cases they have only been made possible due to the generosity and charity of a patron in the form of a vendor or through some other form of sponsorship. As faster RP machines are developed and materials and processes become cheaper, this will make the technology more generally available. The inclusion of colour as an integral aspect of some RP processes is also a very exciting prospect for the artist. With the continued development of more durable materials such as metals and ceramics, along with larger and larger build envelopes, this promises the artist an immense leap forward in the potential application of the technology. There are already rumours afoot that suggest one of the major computer-peripheral vendors is readying an affordable desktop 3D printing device for the market. Whether this is true is not of any concern. Given the simplicity of the general concept of layer manufacture, with an ingenious breakthrough here and there, it is sure to happen soon in some form or other.

The cost of RP is bound to fall and when this reaches a level more affordable to the artist, then I am sure that there will be a massive and sudden demand for this remarkably empowering and creative technology. Early in the Millennium we will certainly see these technologies take their place as primary tools in the artist's studio, facilitating the creation of objects never before thought possible. From the relatively small number of sculptors we have seen using RP to date there have already been some remarkably imaginative and inventive applications of this technology, as indicated in these few examples that follow.

Stewart Dickson (Computers and Sculpture Forum) Artists Statement:

Zoetrope \Zo"e*trope\, n. [Gr. ? life + ? turning, from ? to turn.] An optical toy, in which figures made to revolve on the inside of a cylinder, and viewed through slits in its circumference, appear like a single figure passing through a series of natural motions as if animated or mechanically moved. The proposed subject for this zoetrope is the homotopy or metamorphosis from a simple torus to Costa's Minimal Surface (Fig. 1). The artist proposes to construct a zoetrope (Fig. 2) of a topological metamorphosis from mathematical visual computing. The artist proposes to render the individual 'frames' of the animation in physical materials, in three physical dimensions. The proposed method of execution is to employ computer-aided rapid mechanical prototyping. In this zoetrope, 60 phases of the object which transforms are attached to the edge of a wheel. The rotation of the wheel is 'frozen' using a stroboscopic light, optically synchronized to the 'spokes' of the wheel where the objects project from its edge.

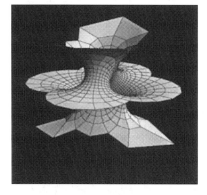

Fig. 1 'Costa's Minimal Surface'   **Fig. 2 'Zoetrope'**

In 'Sirens Trumpet' (Fig. 3) made on the CALM project Helena Swatton, a member of FasT-uk, has taken a computer sound-wave readout of the spoken words 'Come to me' and used this as a contour for a lathe-tool path in a 3D modelling application. She has then output an STL file of the 3D model to be made by the SLA process. This results in the sound of the artist's voice being made manifest in a physical object that takes the form of a trumpet.

Fig. 3 'Sirens Trumpet'   **Fig. 4 'Berlin London'**

In 'Berlin London' (Fig. 4) James Hutchinson, also a member of FasT-uk, has taken satellite photographs of Berlin and London from the Internet and filtered the geography of these two capital cities through various computer programmes and represented them intersected within a single transparent object. This piece was also produced using the SLA technique. From these few pioneering examples it is clear that the combination of microcomputing and layer manufacture offer completely new possibilities for the sculptor. Not only by providing completely new subject matter giving rise to totally new concepts in art, but also as a revolutionary way of making their ideas and creations manifest in a material form.

3   TOWARDS A MANIFEST FORM: FROM SLS TO CAST BRONZE

As a participant in the CALM project, the main challenge was to satisfy the criteria to design a sculpture that would take advantage of the unique features of RP. I decided to design an object that was completely dependent on 3D computer modelling in its conception, with the traditional lost wax bronze casting technique in mind for the realization of the final sculptural work. This was initially because I did not consider any of the RP materials as having any particular aesthetic qualities but saw them simply as a means to another end. Having worked with 3D computer visualization for over a decade, the CALM project provided my first opportunity to produce a sculpture for a 3D output device. It seemed most appropriate not only to take full advantage of the unique qualities of RP but also of those qualities unique to computer modelling; to design an object that could not be realized by any other means. For some years I had been fascinated by the way in which the surfaces of otherwise solid objects offered no resistance to each other when overlapped in the cyberenvironment. Having also been fascinated for some years by the topological peculiarities of the torus (or doughnut), a 3D geometric form having but a single traceable surface and a hole. I chose to manipulate and deform a torus knot, causing its single surface to plunge through and around itself in a complex manner.

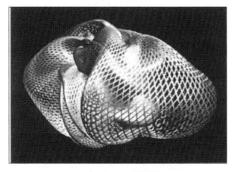

Fig. 5 Wireframe CAD image

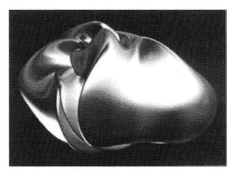

Fig. 6 Solid CAD image

Figure 5 describes the complex way the geometry functions internally. Figure 6 shows the outer surface of the object only. It is impossible to achieve this quality by any means other than computer modelling. The resulting effect is as if modelling something fluid such as a gas or liquid and yet at the same time maintaining complete control over the resulting form as if it were a rigid solid. In places the surface is brought precariously close to an almost knife-edge in thickness. This presented a considerable challenge in making the SLS, the wax, and for pouring bronze. I designed dozens of models before coming close to something that satisfied all of my criteria, since it was part of my initial intention to push the extremities of the SLS technique and lost wax bronze casting process close to their limits. These perceived limits needed to be kept firmly in mind throughout the design process. I chose the SLS process to avoid any post production finishing and to retain, as much as possible, the integrity of the RP process. I received my RP in the post from the Keyworth Institute at the University of Leeds where it had been manufactured, and opened the package with a certain anxiety and anticipation. Seeing it for the first time I was truly amazed by the breathtaking appearance of the object. I couldn't have been more mistaken about the aesthetic qualities of the material. I have been making sculpture for over 35 years, but never have I seen objects that possess the qualities evident in this piece. It was NOT just a prototype as I had expected. The material, DuraForm, in conjunction with the lace-like moiré surface patterns, makes for an immaculate object in itself and not just a prototype. The combined integrity of the concept, process, material, colour, and form added up to something the like of which I had not experienced before; Fig. 7. I thought that I knew this object (inside and out) having spent several days modelling it in the computer, rendering multiple views. It was indeed exactly what I created in the computer but the quality of the object, realized in a manifest material form, completely transcended any preconceptions that I had formed about it as a CAD object.

Fig. 7 SLS 'Continuity of Form'   **Fig. 8 Bronze 'Continuity of Form'**

3.1   Casting the SLS into bronze
Casting the SLS piece into bronze presented a real challenge since I decided to retain as much of the surface texture and integrity of the SLS process as possible. I took several measures to help preserve this integrity. To begin with I inserted a 6 mm mild steel post, with a plate welded on to one end, into the core of the investment through the pouring hole before the investment was set. This was done to provide a stand for the wax so that it could be removed from the rubber mould and worked on without being handled. Simply placing it down on a hard surface might well have been enough to damage the very delicate surface of the wax. The runner and riser system was kept to a minimum, both in quantity and thickness. These were attached around the seam so as to preserve surface detail. The core pins were also placed in the seam for the same reason. Once cast into bronze there still remained the problem of the overall coherence of the sculptures surface. This was somewhat patchy because of feathering caused during the firing process and required further consideration for aesthetic reasons. It is here where artistic sensibilities enter the process with subjective decisions coming into play during the endless hours of careful work that go into chasing and developing the highlights around the complexity of the form to achieve an overall coherence of finish (Fig. 8).

The CALM Project – JISC Technology Applications Programme 314

Elizabeth Hodgson
CALM project manager, Learning Technologies Team, University of Central Lancashire, Preston

1   BACKGROUND OF CALM PROJECT

The CALM project (Creating Art with Layer Manufacture) was set up to encourage the use of the new rapid prototyping techniques for fine art. Rapid prototyping provides a way to produce objects of elaborate topology and great physical complexity, but it is barely known in the art and design community. To set the project in context, it should be noted that previously very few artists had used rapid prototyping at all; perhaps a dozen in all, worldwide. So the CALM project was breaking very new ground in trying to inspire artists to use this technology.

The project was funded by the Higher Education Funding Councils, through their Joint Information Systems Committee, and participation was open to anyone teaching in a UK higher education institution. The CALM project paid for the manufacture of selected computer models, provided by the participating artists.

A Steering Group was set up to guide the project. Art and design is a wide field, and the Steering Group decided that the project should not attempt to reach everyone, but should focus on fine art, as it is the most novel and challenging use of rapid prototyping, and the furthest removed from the usual engineering applications.

The Steering Group also determined the main criterion for selection of proposals: *"the extent to which the proposed objects exploited the unique features of rapid prototyping"*, rather than artistic merit. The intention was to choose proposals that would be challenging to manufacturing as well as interesting to model, and so capture the interest of the engineers involved in the manufacture as well as the artists.

2   ESTABLISHING MANUFACTURING SITES

The first step of the project was the recruitment of a set of sites in UK universities and colleges possessing rapid prototyping equipment, to provide a manufacturing base. Nine sites were recruited, offering the four main manufacturing processes, SLA, SLS, FDM, and LOM:

- Product Design and Development Centre, Queen's University Belfast;
- Innovative Manufacturing Centre, Nottingham;
- Keyworth Institute, University of Leeds;
- Rapid Prototyping and Tooling Centre, University of Warwick;
- Cardiff Rapid Prototyping Centre, University of Wales Cardiff;
- Centre for Rapid Product Development, University of Northumbria at Newcastle;
- Centre for Rapid Design and Manufacture, Buckinghamshire College;
- Rapid Prototyping Unit, Coventry University;
- Product Innovation and Development Centre, University of Liverpool.

The engineers responded enthusiastically to the ideas behind the CALM project, and were very supportive to attempts to introduce RP to a wider audience.

3   RECRUITING ARTISTS

The project was advertised among the art and design community by mailshots and electronic mailing lists, but the most successful method of recruiting artists was word-of-mouth recommendations – people who were interested passed on details to their contacts. Initially, it was supposed that the project would only appeal to those artists already using 3D modelling in their work, but in fact many people were interested who had no background in modelling, or even drawing on computers. About 40 people expressed an interest in joining the project.

4   TRAINING ARTISTS

The next stage of the project was to explain to the artists what would be required by way of a computer model. It was decided to hold a formal one-day training course at Coventry, where the facilities at Coventry University and the University of Warwick would allow the artists to actually see RP machines of different types in action.

The training course was run in Coventry as a one-day event on 22 October 1997. The morning was devoted to presentations on rapid prototyping technology, and demonstrations of several different 3D modelling programs. In the afternoon the participants were first taken to see the FDM machines in the Engineering Department at Coventry University, and then to the Warwick University Advanced Technology Centre, which has SLA, SLS, and LOM machines. Finally the process for making a proposal to the CALM projects was explained, and the proposal forms distributed.

The training material concentrated on those aspects of rapid prototyping considered to be of most relevance to artists, so matters like material properties and construction tolerances were dealt with only very briefly, and more time was given to modelling requirements and techniques. Considerable emphasis was given to the differences between surface and solid models, and the importance of uniting parts together properly in multi-part models.

5   SUBMISSION OF PROPOSALS

There were 35 proposals received altogether. To save the artists from having to spend a lot of time constructing computer models that might never be built, they were only asked to submit sketches of their proposed designs. A total of 22 proposals were selected for building. It was encouraging to note at this stage how successful the training course had been in stimulating interest in the project – 26 of the 32 artists who attended the training course went on to make proposals.

The proposals were very varied in nature. A few of the artists intended to make castings from their models, but most intended to keep the rapid prototyped object as the final artwork, so the choice of material was quite important. Some had strong feelings as to which fabrication method should be used, with SLS proving surprisingly popular for its matte, slightly dusty surface texture, unlike any other manufacturing material. Sanding, painting, and other surface finishing techniques were not at all popular – most people wanted their artwork's unusual construction to be immediately obvious.

6   3D COMPUTER MODELLING

A major concern at the start of the project was the availability of suitable software to the potential project users. A user of the CALM project would have to generate a model either in STL format, or in something convertible to STL. Most conventional CAD packages offer the option of exporting files in STL format. However, the professional CAD programs are expensive, and can be very laborious to master, and were in any case unlikely to be available in art schools. It was felt that artists could not be expected to acquire or learn such technically demanding software if they were not already familiar with it.

The alternative was the wide range of 3D modelling programs aimed specifically at the art and design market. These programs are intended to be used mainly for animation and visualization rather than construction. Typically they offer very powerful facilities for rendering images and backgrounds, but give much less attention to features like exact dimensions than straightforward CAD programs. Many potential users of the CALM project were already experienced in using such programs. However, the concern was that the exported STL files might not be suitable for building real objects.

Some of the animation programs had already been used successfully for rapid prototyping, e.g. Alias/Wavefront, Form Z, and DeskArtes. Others, such as 3D Studio MAX, Amapi, and Rhino, listed STL in their available output file formats, but no-one seemed to have actually tested out the files on a rapid prototyping system. A third group, including very popular programs like Softimage, TrueSpace, and Lightwave, did not export STL files, but did export other formats such as OBJ, DXF, and VRML which could be converted to STL fairly readily.

An additional complication was the large number of artists and designers who only had access to Macintosh computers, so that even quite simple operations like copying a file on to a floppy disk could raise compatibility problems. Many artists did not have e-mail, or access to the Internet, which made the handling of large files more difficult.

In the end, the most successful 'artistic' software packages for 3D modelling turned out to be Form Z and 3D Studio MAX, both of which seemed robust in use, and generated adequate STL files. Several of the artists learned 3D modelling from scratch using Form Z.

7   MANUFACTURE OF MODELS

Once artists had completed their 3D computer models, they were asked to send them as STL or DXF files to the project manager for checking. The files were then sent on to the allocated manufacturing sites for formal quotations. Methods used to submit models varied; attaching files to e-mails, CD ROMs, ZIP discs, floppy disks, and FTP were all used. Although e-mail was obviously the cheapest and quickest method, incompatibilities between different systems caused problems with very large files, and not all attached files arrived intact. The sizes of the model files in STL format varied from 40 KB to over 10 MB, but they were mostly around 1–2 MB.

The Steering Group set a cost limit per artist of £1500, including VAT. The figure was obtained simply by dividing the money available by the number of selected proposals, and originally it was expected that artists would prefer to explore the new medium by making several different small items each. However, only one artist chose to make two smaller models. Most artists wanted one object as large as possible. (The exceptions were the jewellers in the project, who habitually work to specific dimensions.) Manufacturing sites were therefore asked to quote what size could they manufacture an object for the cost limit, rather than what price for a specific size.

The manufacturing sites involved made great efforts to accommodate the CALM project, quoting very preferential rates for manufacture and not charging for engineer's time spent translating, scaling, checking, and repairing models. In some cases there was substantial (unpaid) collaboration between artists and engineers on the best structure for models. Most of the models were straightforward to manufacture, though a few caused problems from either the physical size or the complexity of the computer models (some models had unusually large file sizes because of the inexperience of the artists). Several models were made in two parts to get the final size, then glued together. Several of the artists using FDM volunteered to unpick the support structure from their models themselves, out of interest in the process of unveiling their artwork.

8   EXHIBITION OF MODELS

Towards the end of the project, the manufactured models were gathered together, and were exhibited at several sites around the UK. The exhibition was shown at Manchester Metropolitan University, the Edinburgh College of Art, the Exeter School of Art and Design, and Coventry University. The models provoked a lot of comment and interest at all the sites, but the responses of visitors underlined how little RP is known outside the engineering community.

9   OUTCOME

The CALM project as a whole has been very successful in what it set out to do. It has aroused considerable interest within the target group, namely artists and designers working in higher education, and is expected to have a significant impact on the future practise of the participants. The impact of the project will obviously be greatest for those people who were not already using computer modelling, but it has stimulated a new enthusiasm even among the modelling experts.

All 22 of the selected proposals were completed and built, an impressive testimony to both the determination of the artists in learning new skills, and the efforts of the engineers (given freely) in translating and repairing files from non-standard software. The artists were asked to complete a questionnaire at the end of the project, and the results are presented below. The engineers were also asked, informally, for their comments.

The main outcome of the CALM project has been not so much the rapid prototyped models themselves, exciting as they are, as the network of contacts that has been established, and the increased awareness of both the artists and the engineers of how much they have to offer each other in collaborative work.

10   PROBLEM AREAS

The first problem for many artists was simply that 3D modelling is much harder than it looks, and it takes far longer to make an object than expected, especially for a beginner. Most of the artists grossly underestimated the time it would take them to generate their computer models to their satisfaction, and this led to long delays in the submission of models, and subsequent delays in manufacture. Typically people estimated a few days, and it actually took weeks.

A second difficulty was that the artists often did not understand the limitations of the manufacturing processes. Since the design of an art object is arbitrary (unlike an engineering part which must fit into other parts) the CALM models could have been better tailored to avoid construction problems. File sizes were often excessively large because the artists lacked the experience to know what number of elements would be required to give them the necessary surface quality.

From the engineers' point of view, many of the most conceptually interesting objects made in the CALM project were not really comprehensible without a proper understanding of the artist's ideas. So the engineers could not appreciate which features of the model were important, and which were merely incidental, without talking to the artist. For instance, how important was faceting? Which way up should the object go? Did the colour matter? etc.

However, the worst problem area was the size of rapid prototyped models. The high cost of manufacture meant there were strict limits on how large any individual model could be. This was a serious difficulty for many of the artists, especially the sculptors. Only the jewellers were happy to work within the size constraints of the CALM project budget; everyone else would have much preferred their objects to be made bigger, even at the cost of loss of surface quality. In the long run, it might be that other technologies, such as CNC milling, would be more suitable for making sculpture because objects could be made bigger for the same budget, even though there are many more constraints on the possible designs.

11   EXPERIENCE OF ARTISTS IN CALM

At the end of the project, a questionnaire was sent out to all the artists who participated in the project, to discover what aspects had been successful and what should have been organized differently. The artists were also asked to contribute a short passage summing up their experience of involvement in the project, and what impact it might have on their future work.

Nearly everyone in the project had attended the training course, and they had generally found it at least useful, and in some cases inspirational. Most people were happy with the amount of time devoted to RP technology and to viewing the machines in action, but would have welcomed more information about 3D computer modelling. Most people did all their own modelling, despite describing themselves as 'novices' or only 'moderately experienced'. The typical computer model took weeks to generate, but in the end nearly everyone felt their computer model was very close to their original concept.

The majority of artists felt the RP model was what they had expected to get, and those who had dealings with the engineers found them to be very helpful. A couple of people were very disappointed with the end results. Notwithstanding, all bar one said they would definitely use RP again if it was available, and that the project had encouraged them to make more use of 3D modelling in their future work.

12   EXPERIENCE OF ENGINEERS IN CALM

Although the engineers involved in the project were generally intrigued by the idea of making fine art with rapid prototyping, they were often baffled by the actual objects they were asked to make. The emphasis on the integrity of a production process, and the interest in aspects such as colour and surface texture over dimensional accuracy, seemed very alien. Where the artist had supplied any description or rationale for a sculpture as part of the proposal, this was passed on to the manufacturing site when the computer model was sent for a quotation. (Some of the objects really did not make much sense without a description.)

None the less, all the sites worked hard to produce what was requested by the artists, and in some cases they had to spend a lot of (unpaid) time repairing faulty computer models. Because of the inexperience of the artists, models were often unnecessarily large, and consequently took a long time to verify for manufacture, and to calculate the slice files. This put an extra burden on the engineers, compared with conventional industrial models, but on the whole people seemed to feel that the additional work was compensated by the additional interest. The artists generally spoke very highly of the efforts of the engineers at all the sites to interpret their wishes and construct what was required.

Several sites took up the opportunity to make an extra copy of a model for their own display purposes, where they felt that a sculpture was either a particularly exciting object in its own right, or an unusually good demonstration of the capabilities of the rapid prototyping machines.

13 CONCLUSION

The extraordinary potential of 3D computer modelling for art has been largely overlooked by the academic community in the UK. This has partly resulted from a combination of lack of suitable hardware and affordable, user-friendly software, and partly the more serious problem of disinterest in a technology seen as inherently cold and clinical, and even inimical to art.

The opportunity actually to make tangible, solid objects from computer models has won over many academics who previously regarded a computer model as akin to a paper sketch – not 'real' in any meaningful sense, and so not of interest. In the course of constructing a computer model for rapid prototyping, they have learnt how powerful computers can be as aids to visualization, and overcome the feeling that state-of-the-art technology had nothing to offer them. Those who did have previous experience of 3D modelling have been re-invigorated by the extra discipline of constructing a model for fabrication, and actually holding the final result in their hands.

The engineers in the manufacturing sites have been introduced to a new potential user group for rapid prototyping, and indeed a whole new way of looking at models, as entities suffused with significance and meaning in their own right, rather than just temporary examples of possible future objects.

As well as the experience of the individuals who took part, the CALM project has built up a network of contacts between artists and engineers. These will be carried forward into the future by the organization FAST-UK (Fine Art Sculptors and Technology in the UK), set up by the sculptor Keith Brown, as a direct result of his involvement in the CALM project.

Reverse engineering meets the arts

Chris Lawrie
Delcam plc, Birmingham, UK

1 INTRODUCTION

As reverse engineering and rapid prototyping technologies have advanced over the past ten years, the opportunities for sculptors and artists to take advantage of new media and processes for sculptures and other pieces of art have increased dramatically. Traditionally engineering has been associated with the creation of bland prismatic shapes constrained by manufacturing processes and inflexible design tools. Rarely in the past have engineering organizations and the art world had a need to work together. However, new scanning technologies combined with advanced reverse engineering software tools have offered new opportunities for both artists and engineers, allowing even the most complex of sculpted organic shapes to be prepared for tooling and ultimately mass produced.

Free-form organic surfaces created by the hands of talented artists, can be captured by scanners, manipulated, triangulated, or surfaced by reverse engineering tools before entering into a number of different end applications. Sculptures can become mass-produced miniatures, virtual reality objects, styled consumer items, or simply copies in varying materials. Rapid prototyping technologies have helped to add to these choices.

Detailed below are two case studies, which aim to show just two examples where reverse engineering technologies have expanded the boundaries of art. The first case study shows how reverse engineering technologies can be used to create giant rapid prototype sculptures from miniature originals without losing detail and accuracy. The second case study illustrates how reverse engineering can be used to assist in the creation of high-volume consumer items from fragile and valuable sculptures.

2   CASE STUDY 1: THE MILLENNIUM SCULPTURE

In 1998, Williamson Park in Lancaster commissioned historical sculptor Anthony Padgett to undertake a project which aims to create a snapshot of advanced digital technologies in the form of a 2 metre high mythological sculpture in bronze. The sculpture will stand as a milestone to offer future generations a historical view of rapid technologies at the turn of the millennium. The completed sculpture, which will eventually take pride of place within Birmingham's National Exhibition Centre (NEC), started life as a small 400 mm high model created by Anthony Padgett and takes its influences from art, spirituality, history, nature, and technology.

Fig. 1 The original sculpture

The original scale model was cut into eight pieces, the aim being to use varied reverse engineering and rapid manufacturing technologies to create the full scale giant. At the start of the project 12 companies committed their resources, materials, and technologies to produce individual portions of the sculpture. Delcam were approached by Anthony Padgett late in the project when two of the original companies were struggling to generate detailed full-size models from the scanned data.

2.1   The process

Fig. 2 Body set up for scanning on the Cyclone

Seven of the eight pieces were scanned on a contact scanning system called a Cyclone. Produced by Renishaw, the Cyclone 'traces' the physical part using a probe, which captures points at predetermined intervals along the surface with which it is in contact. This system allows all of the fine sculpted detail to be maintained in the digital copy of the physical sculpture.

To give some indication of the detail captured, Fig. 3 shows the resulting scan of the body and shoulders of the sculpture. This three-dimensional (3D) image is created by triangulating (or polygonizing) the points captured by the Cyclone scanner. The software, CopyCAD from Delcam, was used to create the triangle model from the scanned data.

Fig. 3 The triangle model of the Cyclone scan (to view this figure in colour, go to colour section)

In order to do this, the software read two separate scans (top and bottom) and aligned them to form the 3D shape (still as points). The points were then triangulated to create a closed 'surface model' of the sculpture piece. The 3D digital model was then scaled by a factor of six to produce the finished part of the final giant statue. A software tool, Delcam's Trifix, was used to repair the triangulated model to make a watertight object, ready for rapid prototyping. This cleaning process was necessary, as even the smallest inconsistencies within the triangulated model would have created a very expensive yet unacceptable model.

Rover Group used their laminated object modelling (LOM) rapid prototyping machine to create the body and shoulders of the statue. Because of the large size of the model it was split into two. To reduce the weight of the model, the STL file was modified (hollowed out) to give a 25 mm shell. The finished LOM model is shown in Fig. 4.

Fig. 4 The final LOM model

The detail of the statue's head (the eighth piece) was captured using 3D Scanner's ModelMaker laser stripe scanner. This system works by projecting a laser stripe on to the surface of the part and receiving information back through cameras, which enables the system to determine the exact location of a point on the part's surface. Figure 5 shows the system in operation. In this case the ModelMaker is attached to an articulating arm, which gives the scanner freedom of movement while maintaining a high positional accuracy. This type of scanner can produce one scan file for the entire head.

Fig. 5 The ModelMaker system at work (to view this figure in colour, go to colour section)

The point cloud file (X, Y, Z data) from the scan was read into CopyCAD and a triangle model was produced using the 'points to triangles' wizard. (Note: A triangle model is simply a mesh of triangle facets representing the part scanned.) The triangle model was closed (the neck was capped) and scaled to the final size. The detail of the triangle model is very impressive considering the original head sculpture is not much bigger than a golf ball.

Fig. 6 The triangle model of the head (to view this figure in colour, go to colour section)

Fig. 7 The SLA head

Rover group used their stereolithography (SLA) machine to create the complete head in one build. The final head was very impressive both in terms of its size, and the detail reproduced of all the features from the original miniature models. Figure 7 shows the final SLA part. Figure 8 gives some indication of the scale of the rapid prototype models produced. In the words of Anthony Padgett "The final result is outstanding, the original organic detail within the sculpted miniature figure has been maintained in the final giant LOM body and the stereolithography head".

Fig. 8 The head and shoulders

Through close collaboration, Delcam and Rover Group's Prototyping Centre were able to deliver the two complex prototype models within the remaining time to meet the project deadline. Other parts of the statue were completed using LOM and high-speed machining (HSM).

It is true to say that sculptures such as this are the extreme in terms of free-form organic shapes. If the latest scanning technologies, combined with CopyCAD and prototyping techniques, can reverse engineer and manufacture detailed organic forms to such a high level of quality and accuracy, then these technologies certainly meet the requirements for the 'reverse engineering to rapid prototyping' processes within engineering markets.

Fig. 9 The final six-foot-high rapid prototype statue

3   CASE STUDY 2: THE TATE GALLERY PROJECT

The following case study describes the application of art, CAD, and prototyping in the development of one of the products within the 'At Home With Art' project undertaken by the DIY retailer Homebase and London's world famous Tate Gallery, with support from the Arts Council of England. The project concept was to make contemporary art more accessible and widespread by mass producing designs from leading sculptors and selling them through the Homebase chain. Delcam's contribution came in the production of a trowel and fork set, designed by Turner Prize winner Tony Cragg.

3.1   The process

The original models produced by Mr. Cragg were scanned on a Renishaw Cyclone to produce sets of points describing each shape's form. At this stage the physical parts were returned to the Tate Gallery. From this stage onwards the scanned data became the nominal information upon which all subsequent work was based.

Fig. 10 Scanning sculptures on a Cyclone

The scanned points were read into Delcam's CopyCAD and automatically triangulated. A series of sections were taken through the triangle model to create a grid network of construction lines. Smooth surfaces were created over the curve network. The surface model form created became the new nominal data from which a detailed design and production tooling could be created.

The surfaces were transferred into the CAD (PowerSHAPE from Delcam) system for the designs to be finalised by projecting a series of holes into the surfaces making up each handle. However, on viewing the rapid prototype models it was realized that the modified design would not withstand to repeated use, so additional reinforcing ribs were added to the interior.

Fig. 11 Adding engineering detail

A series of rendered images were then prepared so that Mr. Cragg could confirm that his design had been captured exactly by the software. To provide a final check, an STL file was generated for the production of physical prototypes. These prototypes took the form of an LOM model for the fork and a Z corp model for the trowel. The STL file was also imported into an FEA package to confirm that the designs could be moulded successfully.

Fig. 12 The final design concept

Once the designs had been approved, work began on the tooling design. As well as the core and cavity design, a series of electrodes had to be produced to form the shape of the reinforcing ribs. Data for all the components were processed to generate the tool paths using PowerMILL from Delcam.

The tool paths were sent to Format Precision Engineering for machining of the moulds. These were delivered to AM Tech Services for moulding of the handles. Finally, the handles were sent to Spear and Jackson for assembly into the finished fork and trowel.

All of the work was undertaken with strict deadlines, as Homebase needed 2 500 of each implement for the national launch of the 'At Home With Art' range. Despite the complexity of the design, the involvement of Delcam staff and the Power Solution software meant that delivery was achieved on time. The project time scale from receiving the sculptures to delivering parts to the store was just over two months.

Fig. 13 The finished garden tools

4   SUMMARY

The case studies described here are just two examples of many projects where reverse engineering and rapid prototyping play a vital role in 'art to commercial part'. In today's market place artists are looking to create a large return on their talent and effort within very short time scales. Hence technologies such as reverse engineering and rapid prototyping, which offer artists a new medium to work in while offering a fast track to finished product, are in increased demand.

Concept modelling developments in architecture

Charles Walter Hugh Overy
Director of Engineering, Laser Graphic Manufacturing, Minturn, Colorado, USA

ABSTRACT

Rapid prototyping technologies are beginning to revolutionize the design and production of a wide variety of physical objects ranging from cell phones to statues. It would appear that rapid prototype (RP) technologies should be translatable to architectural design and construction. In particular, the new class of 'concept modellers', intended to provide the designer with an office friendly way to produce physical models, should offer a welcome replacement to traditional methods of model building and visualization. While RP technologies have tremendous potential for architecture and related fields, the rate of adoption has, to date, been slow. An examination of the similarities and differences between the architectural and mechanical/industrial design processes reveal that there are critical differences in the design paradigms, process flows, and market demands. Architects, RP equipment manufacturers, and others experimenting with or studying the use of RP in architecture need to understand both the technical and market requirements that will make the adoption of this technology successful.

1 THE PROBLEM OF COMMUNICATING DESIGN

> *"By the simplest definition, architecture is the design of buildings, executed by architects. However, it is more. It is the expression of thought in building. It is not simply construction, the piling of stones or the spanning of spaces with steel girders. It is the intelligent creation of forms and spaces that in themselves express an idea."* [1]

The difficulty of communicating complex three-dimensional ideas from two-dimensional data has plagued designers of all types for thousands of years. Architecture is one of the oldest of the design professions and its practitioners have been struggling with reconciling their visions with the desires of patrons and clients throughout recorded history. While the builders would have had little trouble explaining the design of the great pyramid, the subsequent forms of complex domed temples, state buildings adorned with columnar arcades, and fortifications with roof overhangs and truss-work, must have been extremely difficult to visualize when drawn in the sand, or frames of hardening mortar. Perhaps, as appears to happen so many times in this age, the client, the neighbours, or the community, nodded their collective heads without any comprehension of what the vast scrawl of lines and annotations would really look like once built. The architect proceeds with the project on the strength of their vision alone, with results that are sometimes what everyone expected, and sometimes what no one expected!

Traditionally, architects have built small-scale models to communicate the general form or concept and to work out design issues that are otherwise difficult to visualize. This is very much the same reason that any designer creates a mock-up or prototype. Architectural models are commonly built by junior staff at a small firm and are often built from simple and

Fig. 1 Concept model of a multi-unit development[2]

inexpensive materials including chipboard and balsa wood or even manila folders and cardboard. If a more precise or elaborate model is necessary for large client presentations, public comment, or sales, a professional model shop is employed. In a large firm this might be a division of the firm itself.

The relatively recent introduction of a class of rapid prototype machines generally referred to as 'concept modellers' offers to alter significantly the production of architectural models. Eventually these technologies will have implications for the architectural design process itself. The term concept modeller is intended as much to describe the use of a machine as it describes the technology itself.[3] I imply a machine that is capable of rapidly and autonomously building a three-dimensional object from computer data, a machine that uses materials which are usually unrelated to the material in the end product and a machine that can reside in an office as opposed to a production environment. At present, the commercially available technologies that best fit this description include 3-D Systems – Thermojet, Stratasys-Prodigy, and Z Corp. Technologies from Sanders and others might be appropriate in specific cases, and topographic landforms can be readily built with conventional computer aided machining (CAM). In addition, there are several other concepts under development or near release[4].

The drivers for the development of these technologies have not been the needs of the architectural community. Rather, the majority of the current successful applications, as well as the vast majority of experience and installed equipment is in the arena of mechanical and industrial design. As these technologies have become more prevalent and the software to generate data to build models has become more sophisticated, first adopters in the architectural community have become interested in the applications of concept modelling. In addition, the equipment manufacturers, academic institutions, and service bureaus are looking outside their existing customer base for new markets. Finally, the small number of professional model shops, of which my company is an example, are seeing a radically new way of doing business that demands totally new business models. Central to the successful marriage between architecture and rapid prototyping is an understanding of the similarities and differences between what I will generically refer to as 'industrial design' and what we, from the perspective of a service bureau, see as the current norm in 'architectural design'. Once these are understood, one can begin to understand the current applications and future directions of this partnership.

I will make reference to two recent cases, the 'Ell Residence' and the 'Log Home', both high-end single family second homes that are currently being built in Colorado. The Ell Residence was a very unique design by a prestigious architecture firm. Four complete models were built for this house, two by our firm and two by the architecture firm. The models were an integral part of the design and review process and also assisted in the solution of complex roof forms. The modelling process was iterative and collaborative using CAD files. In contrast the Log Home model is a more modest project that we originally built using flat styrene parts cut by our high-powered laser cutter. The parts were drawn in a 2D drafting program using digital scans of the hand-drawn plans as a starting point. Additionally, the Log Home is important as, after the fact, we redrew the project using a NURBS based solid modelling program[5] and have samples from a range of concept modellers. We are therefore able to compare the physical models as well as estimate the financial and process implications.

2   ARCHITECTURE AND INDUSTRIAL DESIGN

Industrial design and architecture are not carried out in an artistic vacuum. More often than not they are business endeavours that are driven by business imperatives. Excellent architecture can take place without modern technology. When we look at the adoption of a new and appropriate technology within an existing profession it is unrealistic to consider the benefits to the designer without looking at the economics as well.

2.1   Similarities
Rapid Prototyping technologies have begun to change industrial design because they offer significant economies in a variety of processes. The limited successes that the concept modellers have had in architecture are a result of applications that take advantage of the similarities between the business necessities of these two design paradigms. First, concept modellers offer the designer a clear and unambiguous way to communicate. A model is intuitive to most people especially if it incorporates some elements that give a sense of relative scale. A model is also easy to view in a group, as it does not rely on the presenter to move through the design. Each person is able to review the design in a way that is consistent with their particular their interests, concerns, and background. This attribute of physical

models is extremely important to architecture because preliminary designs are often viewed by clients or by members of the public who have no training in visualizing 3D forms.

Second, most businesses operate within a competitive environment. The use of concept modellers allows a designer a new and impressive medium in which to communicate their design. In addition concept modellers allow the designer more control over the schedule, accuracy and look of the resulting product. Particularly in the current market, new, high-tech methods help to sell innovative or premium products.

Third, concept modellers allow for the visualization of complex forms and design problems that may be difficult to solve in two dimensions. We find with our clients that the intersection of complex pitched roof planes is particularly difficult to perceive and often requires the laborious drafting of multiple building sections. As a result, difficult conditions often discourage experimentation or cause solutions to be left until late in the design process. In contrast, concept modelling and the underlying 3D solid model encourages a variety of solutions because the 2D views, whether in plan or section, are implicit in the model and the mathematics of the plane intersections are left to the computer.

The first and third similarities, the need to communicate and the need to visualize, drove the four iterations of the Ell Residence models. The models that the firm produced in house were attempts to figure out a roof plan that was feasible and aesthetically pleasing. Our models assisted in the roof problem but were further driven by local ordinance that required a model to communicate the design to a review board. In contrast, the Log Home model that we initially produced for the client was solely the result of a design review requirement. In fact the model requirement was seen as unnecessary by the architect and a burden by the developer and client.

2.2   Form, fit, and function

If form, fit, and function are the fundamentals of design, all of the similarities in the applicability of concept modellers to architecture occur under the heading of form. In evaluating the technologies as a method to communicate form, all of the rapid prototyping technologies provide results that are more than adequate. Absolute accuracy is not that important because relatively rough forms communicate the design intent. In our market research we constructed a very coarse, hand-assembled model of the Log Home. The model was built from .1" horizontal slices of the NURBS model that were laser cut and stacked by hand. There was no finish applied and the model took about two hours to complete. We then placed the coarse model beside RP produced models and solicited feedback from our customers. When an architect who had never seen the project before, looked at both models he saw little difference between the two methods for communicating the overall design intent of the building. This reinforces the idea that the present design or marketing of concept modeller technologies is not driven by architectural needs and that market opportunities may indeed exist.

Although concept modeller output satisfies the form requirements, it does not fare well in fit and function tests. While it can be reasonably argued that concept modellers are not supposed to check fit and function, the problems run a little deeper. The most obvious difference in mechanical models and architectural models is scale. Currently, many industrial design parts are prototyped at the scale of the final part. Architectural models are usually built at around one hundredth of actual size or smaller. When the model is scaled down this much, what can be built by the modeller is no longer an accurate representation of what needs to be built by the builder. In the emerging industrial or mechanical design paradigm the computer based, virtual 3D model is validated by the concept model and is also suitable for a variety of computer aided tests such as thermal and stress analysis. In addition, a physical concept model can help in packaging and other 'fit' related tasks. The data for architectural concept models is not as transferable or extensible. A scaled down exterior wall thickness of 8" when modelled at 1" = 8' (a relatively large scale) results in a wall thickness of just .08". While concept modellers will produce detail at this scale, the thin walls create significant problems, including the fact that the model may be too fragile to be useful or even to be built successfully. Smaller elements like a common 2" × 4" wooden truss are .02" × .04" at scale, dimensions that approach the limits of many modellers.

The thin wall problem is further compounded by the most common practices for drafting architectural 3D models. Architectural CAD models are often built by developing a floor plan and then extruding the walls up to the height of the floor above or to the roof. This results in a physical model that has a large volume of spaces enclosed with the very thin walls. In many cases this results in an unbuildable prototype, or at least a model that is far more complex, expensive, and difficult to produce than what is needed. The only reasonable solution for a model shop or a designer outputting to a service bureau is to redraw the model using techniques that will result in file that can be more readily built using the desired technology.

The process of redrawing has little benefit for the architect and results in two files that must be kept current. The NURBS model of the Log Home had no value to our clients. The cost of drafting this model in 3D could only have been recouped on the final cost of the model. If the architect had created the 3D model, it could have been used for sun studies and perhaps for exterior rendering but it would not have been very useful in structural studies or in interior planning. Currently, the labour costs of people skilled in architectural solid modelling and knowledgeable about the requirements of RP processes are much higher than the cost of interns willing to make models in order to get experience in an architectural firm. Add this to the fact that a good CAD station can cost $5000 and an excellent X-acto knife costs $5 and you have created a significant barrier to adoption in most small- and mid-sized firms.

2.3 Process flow

While issues described above are related to the actual model, the differences in the architectural process are equally important. In our experience, architects move to CAD later in the design process than their counterparts in industrial design. All of the drawings for the Log Home, a project with a total cost of over $500 000, were drawn by hand. For the Ell Residence the elevations for the first model were drawn by hand and the floor plans were on CAD. The second Ell Residence model was entirely in 2D CAD. When architects do move sketches into CAD it is not so much to conceive of volumes and to communicate ideas as it is to work on specific details like framing and fenestration. As a result, at the point that architects would like to begin producing a model there is often incomplete 2D data and no 3D data. The necessity of building a model at this level of design can create an artificial barrier in the design process. This is particularly true if the architect must create additional drawings for the sole purpose of creating the model.

Architects also have more opportunity to alter and to finish the design once the project has begun. On one project the designer had no idea how the interior roof forms would be finished until they stood in the room. While the cost of moving a window on a residential project is not cheap, it does not compare to the problems of changing hard tooling during a production run.

There is also a tradition of model making in architecture. In fact, architects build their own models in schools and many continue to use it in their professional careers as part of their design process. Unlike engineering, architects see model building as a skill that should be learned. What we see in our company is that our primary competition is the architectural office itself. There is a high demand for outsourcing if a firm is busy, but if there is a slowdown, model making will be one of the first things to be pulled back in house. This problem can be just as easily translated to the in-house acquisition of expensive equipment.

Finally, while schedules get tight in every industry, the architectural projects have significantly longer total cycle times than many industrial/mechanical designs. The Log Home was under design for about 4 months prior to our concept model and the total design on the Ell Residence will be over a year. Construction will occur simultaneously with some design but building will take over 14 months. These longer cycle times negate some of the benefits of RP.

3 CURRENT APPLICATIONS

Despite the fact that the drivers that led to the development of concept modellers differ from the current needs of the architectural community, applications for RP technology exist today that make sense for the architectural community. These applications are commercially viable, make appropriate use of the technology, and provide a firm foundation for future development.

Currently the technology is a tremendous aid to the model builder if not to the architect or the architectural design process. One of the most clearly effective uses for RP is for multi-unit development projects where many buildings are similar. In many residential and commercial developments cost savings are achieved by taking a basic design and making minor modifications such as changing the angle between building components or mirroring portions of the design. In this case there are few economies using traditional hand-built modelling techniques but the benefits of RP are tremendous. Another application for RP is to model compound curves such as tented structures that are difficult or impossible to model accurately using other methods. RP technologies can also be wonderful for adding accurate or custom details such as facades and trusses that are time consuming and difficult to build using conventional techniques. In all of these special cases, RP is sometimes a wish and sometimes a reality. With the increase in the availability of the technologies it can more often be a reality.

3.1 Terraforms

Fig. 2 Model of a mountain golf course

The construction of landforms is also a good place to begin exploring RP technology. The site survey is one of the first steps in the architectural process. Landform data are usually available as an x, y, z data set, as the surveyors collect it this way. For large areas is may be possible to obtain reasonably accurate data from various government agencies. In addition, there are very few deep cavities, or overhangs in nature, at least not at the scale required in architectural or development models. This combination of factors allows for more traditional technologies like computer aided machining (CAM) to be used in the generation of large, detailed landform models. A simple three-axis mill of average accuracy is capable of producing a variety of useful parts.

4 TOOLS FOR THE FUTURE

In order to create a significant market in architecture, customers and suppliers need to address the shortcomings in both the software and hardware elements of rapid prototyping.

4.1 Software
The importance of the integration between concept modellers and software cannot be underestimated. As I have described, there are serious shortcomings and disconnects when one tries to build concept models using current architectural drafting programs and techniques. Today most architectural CAD packages are heavily oriented toward documenting design and drafting. While some packages offer STL export, there are usually problems if a design of any complexity is simply sent to a service bureau. Instead, the problems and challenges of designing and communicating form must become part of the fundamental utility of architectural software. For example, midrange mechanical CAD applications allow an engineer to see very easily the effect of an altered chamfer radius, or visualize the production issues associated with a mould draft angle. However, to my knowledge, no software utility exists for an architect to experiment, in a couple of clicks, with elements as fundamental to the shape of a structure as the pitch of the roof or the number and type of dormer windows.

In addition, when you have experienced the difficulty and cost of redrawing a design in order to create a concept model, it is easy to conceive of a software function that would create a solid 'skin' around the exterior of an architectural model. This skin would represent a shape that was an inherently buildable STL file. This software enhancement may seem far off but if one considers that the relatively specific function of sheet metal unfolding has recently become the 'must have' feature of mechanical design software, it gives my dreams a sense of reality!

4.2   Total system cost

In many areas of industrial design, RP offers a clear source of competitive advantage and returns on investment have been documented. While the current concept modellers are more than adequate for a wide variety of architectural design uses, I have reviewed why only some of the benefits and cost savings hold up when applied to architectural design and how additional hard costs may be incurred. When these factors are combined with the relatively low cost of traditional modelling techniques and the issues associated with architectural traditions, it is clearer why most architects will not pay a market premium for RP technology.

The total system cost of a concept modeller includes the initial expense of roughly $50 000, plus overhead, additional computer systems, software, and training. Added to this are the variable, per model costs of drafting, materials, and post finishing. When all these expenses are considered, the total system cost is higher than the benefits that can be derived in most cases. Concept modellers are beginning to be sold to some large architecture firms that have the manpower, the capital, and the vision to experiment with new technologies. In addition, RP has been used sporadically for some time on limited projects often by these same companies. What we can expect is as manufacturers of concept models are able to decrease the total cost of their systems, the technology will become appropriate to a larger and larger audience. In addition, as manufacturers become more aware of the unique demands of architectural modelling they will be better able to market and adapt their product offerings. Within the RP community, there have been numerous musings about and rumours of concept modellers with prices below $10 000. I believe that prices below even $20 000 would substantially alter the dynamics of the industry by making the technology widely available to a broad class of users.

5   FUTURE DIRECTIONS

Fig. 3   Concept models of the Log Home. Z-Corp to the left, Stratasys right

It is inevitable that concept modelling will become a part of architectural design. What remains unclear is how soon the technologies will be adopted and who will provide them. The acceptance of 2D CAD throughout the vast majority of the architectural community is only now becoming a reality. In general terms, this is as much as ten years behind mechanical design. However, given the accelerating rate of technology adoption, it is unlikely that the widespread use of RP in architecture will continue to lag far behind other design applications.

ACKNOWLEDGEMENTS

Morter Aker Architects and Zehran and Associates both of Vail, Colorado, USA have provided feedback on a variety of our forays into these new technologies. Lillian Millin and Jason Berghauer of LGM have contributed time and effort to the development of this article. Juile Freeman has provided support and her invaluable proof-reading skills. Z-Corp, Stratasys, 3D Systems, and Sanders have all been very helpful in our explorations of concept modellers. The online community of the Rapid Prototyping Mailing List deserves enormous credit for their continual efforts to expand and share the knowledge of rapid prototyping.

Endnotes:

[1] Excerpted from Compton's Encyclopedia. Copyright © 1994, 1995, 1996, 1997 The Learning Company, Inc. All Rights Reserved.

[2] Bearpaw Lodge at Beaver Creek Resort in Colorado. Design by Morter Aker Architects, Vail, Co. Development by East West Partners. Model by LGM, Minturn Co. Model built from styrene laser cut parts from two dimensional landform and building data.

[3] All of these machines are being used for a wide variety of tasks that do not include the generation of pure concept models. Many of these applications may in fact be integral to the design and construction of buildings such as the design of custom mouldings or the manufacture of special hardware using investment casting. However, these processes are not unique to architecture and are better described elsewhere in this volume.

[4] www.cc.utah.edu/~asn8200/rapid.html contains one of the most up-to-date lists of RP technologies.

Section 6

The Future

Rapid manufacturing

Professor Phill Dickens
Department of Mechanical and Manufacturing Engineering, De Montfort University, UK

ABSTRACT

There has been considerable interest in industry over the new processes known collectively as rapid prototyping. If these processes can be developed into more robust manufacturing processes for making end use parts then they would have a dramatic effect on the design and manufacturing world. There has been some limited use of the existing rapid prototyping processes to manufacture end use parts but there are major limitations that have held this back. There exist several limitations currently such as machine output, machine and material cost, part accuracy, and surface finish.

1   INTRODUCTION

It is likely that rapid prototyping technology will eventually move into the heart of the manufacturing process and end use parts will be made by these additive techniques. The term rapid manufacturing has been coined to reflect this new application.

2   EXISTING USE OF RAPID PROTOTYPING FOR MANUFACTURING

Little research has been performed to assess the commercial feasibility of using rapid prototyping processes to manufacture production parts in large volumes. It has been shown that even if all times and costs were to be reduced to 1 per cent of their current value throughout all the stages of production, rapid prototyping would still not be able to compete with injection moulding in terms of large volume production (1). Generally the cost of producing the parts by rapid prototyping was about 100 times greater than with injection moulding. The production rate with injection moulding was about 100 times greater than with rapid prototyping. However, this work showed that rapid prototyping processes could be used for low volume production (up to 6 500 parts) and possibly for medium volume production in the future.

An example of manufacturing is in the production of small complex electronic housings (up to 30 % 30 % 10 mm) (2). The FBI (USA) also manufacture custom-designed consumer products to enclose surveillance equipment (3). Another unusual application is the manufacture of end use parts for the Spacelab (4). Recently selective laser sintered parts have also been flight approved in the USA (5) and some special parts are being made, such as filters, that would be impossible to manufacture by any other process.

3 CURRENT LIMITATIONS

3.1 Materials
One of the main limitations at the moment is the limited range of materials. With current commercial rapid prototyping machines there are about 80 materials available compared to tens of thousands of polymers for just one process such as injection moulding. The mechanical properties of these materials has improved greatly over the last ten years but it is only recently that they have become more useful for end use applications with better impact strength and flexibility. There is still a long way to go before designers have a range of materials to choose from with these techniques. Rapid prototyping materials are also very expensive compared to conventional engineering materials. For example, stereolithography resin costs about £150 per kg, whereas, a common engineering polymer (ABS) costs about £0.4 per kg.

There needs to be a range of materials but not necessarily on the same machine. These materials will need to compete with those conventionally available but we do not necessarily want to process conventional materials. There is an exciting possibility for a range of completely new materials. These materials need to be fully characterized over the entire range of operating conditions.

3.2 Accuracy and surface roughness
The accuracy possible with rapid prototyping processes is still not as good as many conventional manufacturing processes. In many cases it is difficult to hold tolerances better than +/– 0.1mm. The other main limitation is the poor surface finish that is obtained with many processes.

3.3 Production speed
The previous work (1) was undertaken three years ago and rapid prototyping machines have increased their output considerably since then. For example the recently released SLA7000 is claimed to be four to eight times faster than the SLA5000 which in turn was twice as fast as the SLA500 used three years ago. If this rate of improvement continues then the rapid prototyping processes will start to compete with conventional processes such as injection moulding of plastics.

3.4 Machine cost
Rapid prototyping machines are also very expensive and continue to be so, however, the investment cost itself may not be the problem but the output per £ investment is important.

4 POTENTIAL BENEFITS OF RAPID MANUFACTURING

There are a number of important benefits that could be obtained from rapid manufacturing.

4.1 Tooling
The tooling cost could be greatly reduced or eliminated. This would reduce the risk involved in new products and would encourage greater variation in designs. The elimination of tooling would also reduce lead times for new products.

4.2 Design freedom
One of the most remarkable abilities of rapid prototyping is the ability to manufacture any geometry. This is a radical development in manufacturing because all the conventional processes are severely limited in the geometries they can produce. Manufacturing engineers have spent the last ten or 20 years convincing designers to develop geometries that can be made more easily. We are now close to being able to say 'You design it and we will make it!' It will be possible to have re-entrant shapes without complicating manufacturing, no draft angles, no split lines and completely variable wall thickness. It will also be possible to reduce the number of parts leading to easier assembly and lower parts counts.

4.3 Graded materials
A more recent development that is being investigated is the manufacture of graded materials. With the additive processes it is possible to place different materials in different places and perform the transition from one to another under great control. This has always been difficult with conventional manufacturing processes.

4.4 Flexible manufacturing
If tooling is not being used then it will be possible to change from one product to another very easily without affecting production efficiency. It would be possible to manufacture much more on a just-in-time basis, as there would be no need to produce large quantities to amortize tool changeovers. This would also lead to less storage requirements for parts.

4.5 Sales
With the added flexibility it would be much more economic to manufacture customized products and customers could start to get involved with design modifications.

5 THE FUTURE

There are many questions yet to be answered, for example, how quick do the systems need to be? What price reductions would make the systems comparable with conventional manufacturing techniques? These are the types of questions that we must be addressing to try and assess the likelihood of rapid manufacturing occurring, and if it should, when and how it would be achieved. The existing rapid prototyping machines can be adapted for rapid manufacturing but in the long term new machine designs are required. The main change that must occur is an increase in build speed. In the medium and long term this will need to be reduced to a few minutes to make the processes comparable to conventional manufacturing techniques. Limitations in terms of materials, part accuracy, and surface finish, together with restrictions on machine build sizes will also need to be addressed.

REFERENCES

1   **Dickens, P. M.** and **Keane, J. K.** (1997) Rapid manufacturing. *Proceedings of European Stereolithography User Group Meeting*, 3–5 November 1997, Florence, Italy.

2   **Iveson, N. J.** (1998) Bigger is not always better. 3D Systems North American Stereolithography User Group Meeting, 1–5 March 1998, San Antonio, Texas.

3   **Ayers, K.** and **Hilbrandt, F. J.** (1998) The usage of stereolithographic parts as the final product. 3D Systems North American Stereolithography User Group Meeting, 1–5 March, San Antonio, Texas.

4   **Bampton, C.** (1998) Solid freeform fabrication (rapid prototyping) for liquid fuelled rocket engines. Boeing – Rocketdyne Propulsion and Power Systems, abstract submitted for TCT'98 Conference, Nottingham 1998.

5   **DTM** (1998) Who says the sky's the limit with glass-filled nylon SLS parts. Horizons, Winter 1998, pp 6–8, company literature (DTM Corp., 1611 Headway Circle, Building 2, Austin, Texas 78754).

The rapid production of microstructures

William O'Neill
Micromanufacturing Laboratory, Faculty of Engineering, University of Liverpool, UK

1   INTRODUCTION

During the last 15 years, techniques largely borrowed from the microelectronics industry have advanced sufficiently for the realization of an increasing range of microstructures which can be used as components for microsystems. Early process technology restricted devices to semiconductors and thin metal films through the use of large batch production techniques. The microelectronics industries have engaged in the process of product miniaturization as the resolution limits of manufacturing process technologies have increased. We are all familiar with tiny circuits that power almost every electronic product on the market. More recently we have seen the design freedoms increased for macro products with the advent of advanced machining systems, computer aided design (CAD) packages and rapid prototyping (RP) technologies. The advent of new manufacturing challenges such as those posed by new fields of study such as process intensification, micro fluidics, micro medical devices, sensor laden consumer products and digital telecommunications have generated a considerable demand for new manufacturing techniques to cover a wider range of materials such polymers, metals, ceramics, and composites. A selection of current micro manufacturing techniques is shown in Table 1. A detailed knowledge of these techniques is essential in order to make the right choice for the one that is best suited to produce the required microstructure.

Rapid prototyping systems such as stereolithography (SLA), selective laser sintering (SLS), laminated object manufacturing (LOM), and fused deposition modelling (FDM) have provided substantial benefits terms of reduced lead time and overall cost reduction for macro product development. Micro engineers have yet to benefit from equivalent technologies on the micron scale. This situation is changing slowly as new techniques are being introduced albeit for polymer systems. The objective of this chapter is review the processes and capabilities of the main micro rapid prototyping systems for the rapid production of micro-featured components.

Table 1 Micromanufacturing techniques

Process	Material types	Resolution micron	Batch size	Ref
UV lithography	Poor	< 1 micron	High	[1]
Wet etching	Large	> 1 micron	High	[2]
Dry etching	Medium	> 1 micron	Low/Medium	[3]
Stereolithography	Poor	> 5 microns	Low	[4]
Micro milling	High	> 1 micron	Medium	[5]
Micro-EDM	Medium	> 5 microns	Low	[6]
Laser cutting	High	> 5 microns	Medium	[7]
Localized electro-deposition	Medium	> 10 microns	Low	[8]
Spatial forming	Low	> 10 microns	Medium	[9]
Laser CVD	Medium	> 10 microns	Low	[10]
LiGA	Poor	< 1 micron	Low	[11]
Laser machining	Large	> 5 microns	Medium	[12]

2   PROCESS METHODOLOGIES

So far very little research output has been realized on the micro forming of metallic or ceramic components, in contrast to polymeric materials which have been the subject of a wider range of research studies. This chapter will concentrate primarily on the processes used for the micro fabrication of polymer systems utilizing photolytic polymerization processes. Two main approaches have been used to fabricate three-dimensional microparts, these are direct processes using either vector-vector point curing or integrated planar curing techniques.

2.1   Vector–vector point curing

One of the earliest reported studies on microstereolithography was presented by Ikuta and Hirowatari [4]. Point wise curing to form a solid layer was employed as in macro SLA techniques. Process resolution in terms of minimum voxel size was reported to be 5 x 5 x 3 μm^3. 3D Systems Incorporated introduced the small spot SLA system in 1990s in order to increase the resolution of their SLA250 series machines. The traditional HeCd UV-laser was replaced with a laser capable of delivering low-order spatial modes resulting in smaller spot sizes (and increased resolution in the *xy* plane). The minimum spot size of a small spot system is around 0.08 to 0.1 mm. Despite the benefits of a smaller spot the inherent build methodology of point wise curing and liquid levelling through doctor blades meant that true high-resolution microparts could not be achieved with this process.

The first commercial provider of micro RP services for three-dimensional components was MicroTec of Duisburg, Germany. The process is called rapid micro product development (RMPDTM) [13]. It is a proprietary process and very little is known about its methodology although MicroTec have had considerable success in the development of microsystem technologies by integrating their RPMD processes with complimentary microparts such as jewelled bearings and shafts, etc. A CAD file is sliced in the traditional way and the slice data are sent to the machine slice by slice. The process uses a high-resolution UV laser curing process, which can cure in parallel by splitting the main beam into an array of beams. The component is built up in layers down to 1 μm and a resolution in any direction of up to 10 μm. The minimum component size can be 1 x 10 x 10 μm and is currently limited to a maximum of 50 x 50 x 50 mm. Parallel processing is possible due to the low energy requirements of micro-curing with spot sizes of the order of several micrometres. Parallel construction in this way lends itself to high-volume batch production with economies of scale similar to those employed in the microelectronics industry, Fig. 1 (note: all figures are given at the end of this paper).

With RMPD it is also possible to apply different resolutions during the generation process on the same part. For example, inside a part a coarser resolution is used to accelerate the growth and therefore reduce generation time. At the surface layers, where high surface qualities are desired or finer structures need to be generated, a finer resolution is applied. Figure 2 shows an array of micro-impellers produced in parallel using the MicroTec RPMD process.

MicroTec have further refined their process with the incorporation of a duo-beam system that provides beam dimensions suitable for macro and micro features. Figure 3 shows a CAD model of the smallest submarine in the world and the real thing produced using the RPMD process. MicroTec have also developed process routes for the production of metallic and ceramic microparts using downstream processes. The author is not aware of any plans of MicroTec to sell RPMD systems.

2.2   Integrated planar curing

The integral approach to part building by photopolymerization involves the use of spatial light modulators (SLM) or dynamic mask technology. Dynamic masks are used to provide spatial control over the curing process where each layer is cured at once. The resolution of these systems is dependent on the resolution of the mask and the de-magnification of the imaging system. A number of research groups and industrial developers have used this approach to produce high-resolution 3D micropart building.

A team from the University of Sussex, UK, led by Professor C. Chatwin, has developed an integrated curing system that utilizes a spatial light modulator based on a LCD mask (800 x 600 pixels) capable of transmitting UV wavelengths >350 nm [14,15,16]. Figure 4 shows the elements of the system. An argon ion laser operating at 351.1 nm is used as the light source. In order to provide uniform illumination of the SLM, a diffractive optic element transforms the initial Gaussian beam. A multi-element imaging lens is used to de-magnify the output from the SLM and illuminate the surface of the resin bath. The build platform is controlled by a high-precision motor drive, which lowers the platform into the tank as the layers are processed. No doctor blade or other levelling technique is applied; instead, a deep dipping routine is used to control liquid layer thickness. Figure 5 shows the prototype micro SLA system. Early builds used layer thickness between 35 and 50 microns with build times of 30–60 s per layer, Fig. 6. The next version of the Micro SLA system will use a layer thickness of 5 microns with increased build speeds.

The systems discussed thus far have all relied upon UV curing of liquid monomers. Operating at UV wavelengths is expensive since the cost of UV photons is high, as are the optical systems necessary to manipulate them. Until recently the ability reliably to cure liquid polymeric monomers using visible light was thought impossible if not impractical. Advances in photo initiator developments and resin chemistry have produced stable resins that can be cured at visible wavelengths. These developments have lead to the production of visible-light-cured parts on the macro and micro scale. There is growing activity in this area and considerable progress has already been made. Efforts in this direction are being encouraged by the possibility of creating extremely low-cost RP systems that can process both macro and micro components.

Table 2 Specifications for Envision Technologies 'Perfactory' Macro/Micro RP System

Hard Wired Control Unit
A Multimedia Projection System based on DLP technology from Texas Instruments
Visible Light Curing
Cartridge Material Supply
Photopolymerization basement with a transparent base, working as a reference and polymerization surface.
Photopolymerization basement with a transparent base, working as a reference and polymerization surface.
Build times are quoted as being < 1h/50mm (100µm a layer) independent from part size and structure
Standard build: XYZ (256x192x250mm): Res in XY (100 dpi): Res in Z > 300 dpi
Micro build: XYZ (26x20x30mm): Res in XY (1000 dpi): Res in Z > 1000 dpi

One such company which plans to launch its products early 2001 is Envision Technologies Gmbh, Germany. Envision Technologies plans to bring an affordable, modular desktop RP system 'Perfactory' into the market with a target price of around 22.500 € and 75 €/kg for the standard material. The system is targeted at all creative activities such as product designers, jewellers, industrial designers, model builders, artists, etc. In addition to macro component scales, a special application of the device will be targeted at microsystem technologies such as those used in MEMS industries and medical device technologies. One basic process/device is used as a platform to supply the whole market with simple modular components allowing change from the macro scale to the micro scale, Fig. 7. The system has an imbedded PC for data processing and imbedded Internet interfaces for easy communication with networks. Operation of the device is done through the user-PC connected to the network and can be executed in auto- and expert mode. Process and materials are developed in-house with the key features described in Table 2.

Other system properties include enhanced resolution [factor 2 in the build layer (XY-surface)], with an additional device extension and software update. A propriety transmission control allows the user to control the thickness (over- and under-curing) in selective areas of one layer. Process materials will include methacrylates, epoxydes, ormocers, nanoparticle filled materials and compounds. Volume shrinkage of the material is said to vary from below 1 per cent up to 7 per cent. Envision Technologies are engaged in continuous in-house developments such as those necessary for highly viscous and highly filled materials. Figure 8 shows a selection of microparts constructed with the prototype system

Visible light curing systems have also been developed by a research team at EPFL Switzerland, lead by Dr Arnaud Bertsch [17, 18, 19, 20, 21, 22, 23]. The system used at EPFL is very similar to that employed by Envision Technologies except that a conventional bath with free liquid surface is employed with a transmissive LCD spatial light modulator Fig. 9 and Fig. 10. The machine is a table top system which builds in 5 µm layers with an exposure time of 1 s per layer. The recoating process employs deep dipping and takes 10–15 s to complete. Fabrication speeds of 1–1.5 mm per hour can be obtained with 200–300 layers per hour. Typical microparts built with this system have had 50 to 1500 layers in total. Due to the design of the prototype system the build area is restricted to 3 x 2 mm². Specific photopolymerizable resins sensitive to visible light have been developed using acrylate monomers and free radical polymerization. Due to the leakage of light through the system high levels of absorptivity are required in order to control the cure depth. This occurs naturally with UV light used in the traditional process. The process requires blackout conditions to work effectively and good-quality illumination from the SLM such that dark fields can be kept as low as possible. The system is extremely capable as can be seen from the images in Fig. 11. By comparison, the development of the small spot SLA process has not provided sufficient accuracy for it to be widely used by the MEMS community, Fig. 12.

3   SUMMARY

Microengineering developments are set to account for a substantial fraction of the manufacturing activity in the coming years. These developments will bring a substantial demand for micro prototyping systems but an even greater demand for micro production systems. The batch production of microparts produced by Micro Tec is a good example of what can be achieved with the technology. There are of course significant barriers to progress; the properties of visible-light-cured polymers are below those of conventional UV-cured polymers. Downstream reproduction techniques such as injection moulding and embossing are still required for real functional plastic parts such as glass-filled nylon, PTFE, ABS, etc. The long-term properties and performance of visible-light-cured polymers have yet to be established, although they hold significant promise for low-cost, efficient, and accurate systems. What research remains to be done in this area? Apart from refinement of present systems, I would argue that metals and ceramic technology is essential. Once we have the ability to create accurate micro-metal and micro-ceramic parts, the MEMS industry will be given even greater design freedoms that provide products that will impact on all our lives even more so that it is at present.

ACKNOWLEDGEMENTS

I would like to take this opportunity to thank the contributors to this work: Professor C. Chatwin, University of Sussex; Dr Andrea Reinhardt, Micro Tec Gmbh; Dr Arnaud Bertsch, EPFL; and Dr Hendrik John, Envision Technologies. Without their sterling efforts, the technology of micro rapid prototyping systems would be a pipe dream.

REFERENCES

1   **Despont, M.,** *et al.* (1997) High aspect ratio, ultrathick, negative-tone near UV photoresist for MEMS applications. *Proc. IEEE MEMS*, Nagoya.

2   **Lee, D. B.** (1969) Anisotropic etching of silicon. *J. Appl. Physics*, **40**, 4569–4574.

3   **Jansen, H.,** *et al.* (1996) A survey on the reactive ion etching of silicon in microtechnology. *J. Micro-mech. Microengng*, **6**, 14–28.

4   **Ikuta, K.** and **Hirowatari, K.** (1993) Real three dimensional micro fabrication using stereolithography and metal moulding. Proc. IEEE MEMS, 42–47, Fort Lauderdale.

5   **Friedrich, C. R.** and **Vasile, M. J.** (1996) Development of the micromilling process for high aspect ratio microstructures. *J. Microelectronic Systems*, **5**, 33–38.

6   **van Osenbruggen, C.** (1969) High-precision spark machining. *Phillips Tech. Rev.*, **30**, 195–208.

7   **Chang, J. J.**, *et al.* (1998) Precision micromachining with green lasers. *J. Laser App.*, **10** (8), 285–290.

8   **Madden, J. D.** and **Hunter, I. W.** (1996) Three dimensional microfabrication by localised electrode position. *J. Microelectromech. Syst.*, **5**, 24–32.

9   **Taylor, C. S.**, *et al.* (1995) Spatial forming a three dimensional printing process. *Proc. IEEE MEMS*, 203–208, Amsterdam.

10   **Johanson, S.**, *et al.* (1992) Microfabrication of three-dimensional structures by laser chemical processing. *J. Appl. Physics*, **72**, 5966–63.

11   **Becker, E.W.**, *et al.* (1986) Fabrication of microstructures with high aspect ratios and great structural heights by synchrotron radiation lithography, galvanoforming, and plastic moulding (LIGA process). *Microelectron. Engng*, **4**, 35–56.

12   **Ogura, G.**, *et al.* (1998) Contract manufacturers take on laser micromachining. *Laser Focus World*, pp. 213–226, May.

13   **Reinhardt, A.** and **Götzen, R.** (1999) Microstructure and systems production with Rapid Micro Product Development (RMPD). *Proc. Time Compression Technologies*, Nottingham, 12–13 October.

14   **Huang, S., Heywood, M. J., Young, R.C.D., Fasari, M.**, and **Chatwin, C. R.** (1998) *Microprocessors Microsystems*, 22, p. 67.

15   **Chatwin, C., Farsari, M., Huang, S., Heywood, M., Birch, P., Young, R.** and **Richardson, J.** (1998) *Appl. Opt.*, **37**, p. 7514.

16   **Fasari, M., Huang, S., Young, R. C. D., Heywood, M. I., Morell, P. J. B.**, and **Chatwin, C. R.** (1998) *Opt. Eng.*, **37**, p. 2574.

17   **Beluze, L., Bertsch, A.** and **Renaud, P.** (1999) Microstereolithography: a new process to build complex 3D objects. Symposium on *Design, Test and microfabrication of MEMs/MOEMs*, Proc. SPIE, **3680** (2), p. 808.

18   **Bernard, A.** and **Taillandier, G.** (Eds) (1998) *Le Prototypage Rapide* (Hermes, Paris).

19   **Bernhard, P.** (1997) Proform devises method for adding a second small-spot laser to its SLA-250/40. Rapid Prototyping Report, December (CAD/CAM Publishing).

20   **Bernhard, P.** (1998) *Proceedings NASUG Meeting*, 1–5 March, San Antonio, Texas, USA.

21   **Bernhard, P., Hofmann, M., Schulthess, A.** and **Steinmann,B.** (1994) Taking lithography to the third dimension. *Chimia*, **48** (9), pp. 427-30, 1994.

22   Bertsch, A., Lorenz, H. and **Renaud, P.** (1999) 3D microfabrication by combining microstereolithography and thick resist UV lithography. *Sensors and Actuators: A,* **73**, 14–23.

23   Bertsch, A., Zissi, S., Jézéquel, J. Y., Corbel, S., and **André, J. C.** (1997) Microstereophotolithography using a liquid crystal display as dynamic mask-generator. *Micro. Tech.,* **3** (2), 42–47.

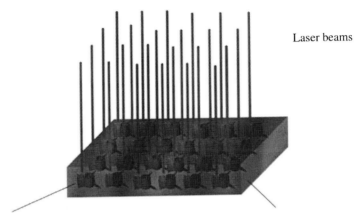

Laser beams

Micro-structures
after polymerization

Light curable acrylate or epoxy

**Fig. 1 Light curing with parallel laser beams
(courtesy of Microtec Gmbh)**

300μm

**Fig. 2 Impellers in parallel production
(courtesy of Microtec Gmbh)**

Technical data:
Shaft diameter: 10 µm
Screw diameter: 600 µm
Hull diameter: 650 µm
Total length: 4 mm

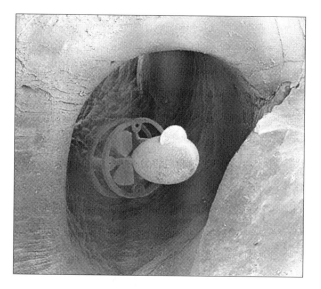

**Fig. 3 Micro submarine CAD image and the micropart seen inside a human artery
(courtesy of Microtec Gmbh, to view this figure in colour, go to colour section)**

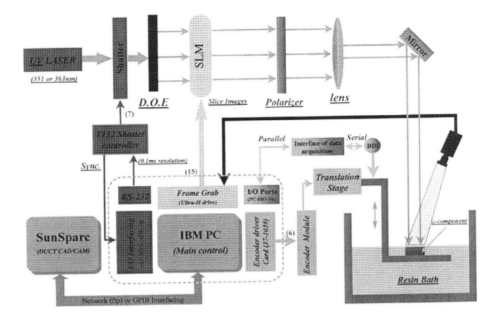

Fig. 4 Micro SLA process system schematic
(courtesy of Professor C. Chatwin, University of Sussex)

Fig. 5 Micro SLA protoype system
(courtesy of Professor C. Chatwin, University of Sussex)

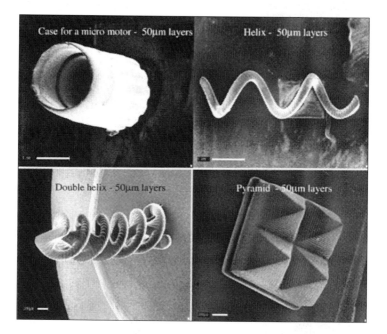

Fig. 6 Micro parts built using the Micro SLA system
(courtesy of Professor C. Chatwin, University of Sussex)

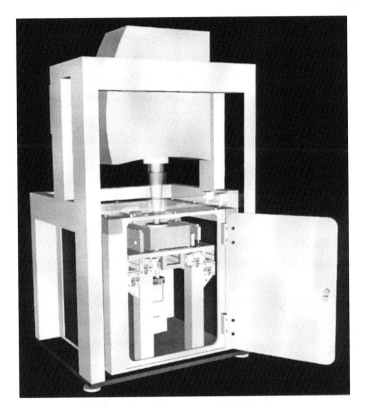

**Fig. 7 Envision Technologies 'PerFactory' Macro and Micro RP system
(courtesy of Dr Hendrik John, Envision Technologies)**

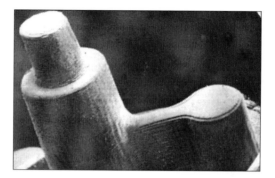

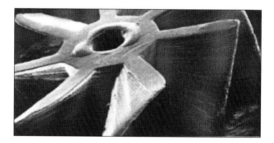

**Fig. 8 Micro components built using a prototpye 'Perfactory' from Envision Technologies
(courtesy of Dr Hendrik John, Envision Technologies)**

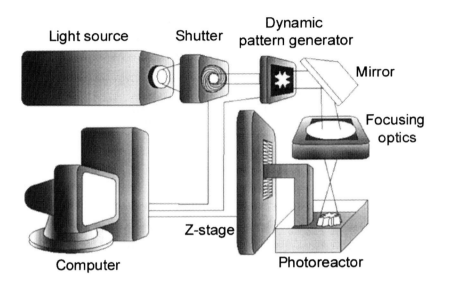

**Fig. 9 System schematic for the EPFL micro RP system
(courtesy of Dr Arnaud Bertsch, EPFL)**

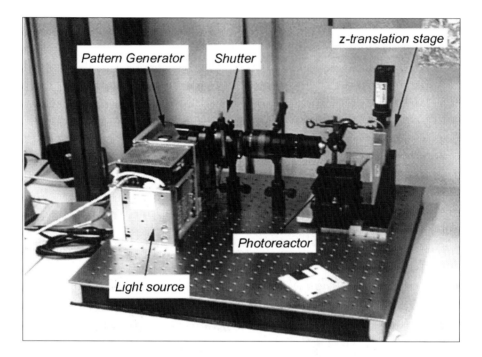

Fig. 10 EPFL micro RP system
(courtesy of Dr Arnaud Bertsch, EPFL)

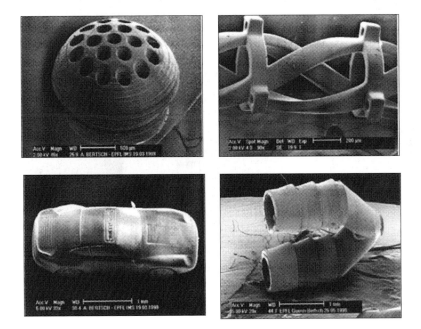

**Fig. 11 EPFL micro-parts; clockwise from the top left; hearing aid; interwoven spring;
double edge micro-fluidic connector; Model Porche
(courtesy of Dr Arnaud Bertsch, EPFL)**

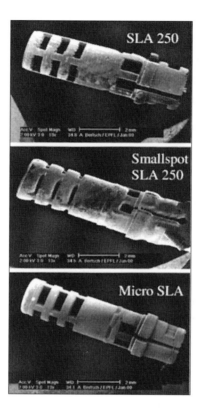

**Fig. 12 EPFL micro-parts; showing a comparison between an SLA 250,
a Smallspot 250 and the EPFL micro SLA system
(courtesy of Dr Arnaud Bertsch, EPFL)**

Index